Loving Math

Advanced Problems with Solutions, Applications and Comments

by

Lewis Forsheit

 www.trafford.com

North America & international
toll-free: 1 888 232 4444 (USA & Canada)
fax: 812 355 4082

Table of Contents

Problems

Geometry and Trigonometry

Algebra

Statistics

Calculus

Solutions, Applications and Comments

Introduction

This book was written for high school students (as I was) and high school teachers (as I am) who love exploring beyond standard math curricula for a deeper understanding of the principles and applications of mathematics. That is what makes for an "engaged student." This also includes anyone who still loves the pursuit of a problem solution, including both professional and amateur mathematicians. The vehicle here that transports us through this exploration is the study and solution of classical and advanced math problems.

As a high school math student, an engineer, a businessman and, ultimately, a high school math teacher, I collected and created math problems and solutions that can be used for advanced study. Some of the problems may be very familiar to you; some may not. A few may be quite easy to do; others will take more time. Included are classical proofs and their extensions that are often omitted in today's curricula.

Beyond the pure enjoyment of this exploration, what do we understand "deepening understanding" to mean? This can mean something very different to each person who experiences it. For some students it may mean the ability to solve really difficult problems more quickly and accurately. Let me suggest that there are four larger aspects of "understanding" and that this text attempts to address each of them.

This first is the aspect of convention. What are the tools we use? What are the vocabulary, terminology and symbols we employ to communicate with each other? How do we best organize the problems and solutions we address to get to the underlying principles we study? The problems and solutions presented here are in a formal style for greatest clarity.

The second is the aspect of evidence. As mathematics is arguably the most rigorous and precise of all human studies and pursuits, evidence must be presented that withstands scrutiny in perpetuity. Algebraic manipulation and symbolic reasoning are used extensively as a core element of the evidence presented. The student will develop this skill considerably by working through the problems.

The third aspect is that of perspective. How do we approach a given problem? Are there different ways of solving it? Can different tools be used toward the same end? How can we develop our thinking and learning skills to address new problems? How can we develop our ability to "think about thinking," which can be seen as the ultimate endpoint of education? Many of the problems here are presented with several solutions to affirm both our ability to learn principles from different approaches and to illustrate that the unity of the mathematics itself is self-consistent.

The last, and perhaps, most important is the aspect of connection. How do all the pieces fit together? How are geometry, algebra, trigonometry, statistics and calculus integrated and inextricably linked? How does the math itself get taught? How does it fit in with other subject disciplines? How is it applied in the "rest of the world," (as if there were such a place)? Might we even ask how understanding deeply allows us to contribute to society and work well with others while doing so?

A portion of these aspects is addressed in the solutions, themselves. The rest is in comments, which come after the solutions. But, these comments only scratch the surface and this book is an invitation to carry on the dialog.

It is hoped that this book will assist teachers and students alike in exploring the subject of mathematics in a new way, whether using material that is thousands of years old, or recently developed. Each problem can be used as a single assignment, done in a few minutes, or a term project that could require intuition, technique, research and/or fortitude (to plow through it). The material can be adapted for use in the standard classroom, subject to students' ability and the constrictions of uniform curricula. It is, perhaps, more applicable to classrooms with the freedom to experiment with project learning and with longer assignment periods. School math clubs or math teams might find this text a handy reference to hone skills, learn new techniques and satisfy the quest for more exciting material beyond the routine.

Although the primary focus here is the application of math principles to math problems, these studies are extended to interdisciplinary explorations of the sciences, engineering, finance, social studies, etc. Examples of these extensions are given and teachers are encouraged to be creative in using this material with students.

The subject material itself is organized into groups. There are twenty-two geometry/trigonometry problems, many of which are "classic proofs." Though some have been forgotten or ignored at large, they are offered here with some new ideas and approaches. There are ten algebra problems, all of which are extensions of a standard curriculum, and offer fresh insights when studied as a group. Statistics, the newest subject to be added to the high school curriculum, has three problems. And calculus, which is not always studied in high schools, has five problems.

Three topics are presented as both a specific case and a general case problem and, as such, include some repetition. Because teachers disagree as to whether it is better to start with a specific case or a general case, the teacher is given the option here. Also, in several instances, two, three or four very different solutions are shown in the exploration of a math principle. In this way, interrelationships as well as creative solution approach skills are developed.

Certainly, there are other problems, not included here, that are just as interesting, if not more so. Also, assuredly, there are more elegant solutions than the ones that are offered. Finding these should only increase the satisfaction and the reward that awaits the intrepid math student at the victorious problem conclusion and "QED."

The text is designed as a workbook with Part I presenting the problem statements with room for the student to work out solutions. Part II restates the problems and offers one or more solutions with applications and comments. The comments range widely, including: additional points regarding the math itself, historical factoids, linguistics, suggestions for teachers, some personal experiences regarding the material, etc. Readers who only skim the problems and solutions might still find the applications and comments quite interesting.

There is a flow to the problems and some use principles from those that come earlier. It is recommended that within a given problem, parts A, B etc ought to be done in sequence (except, perhaps, for those cases in which the "general case" is selected first). Still, it is not necessary to do the problems in sequence; almost any order will work well.

Not only are deeper understanding, skill development, mental acuity, a sense of accomplishment and having fun prospective outcomes for the reader. May we also suggest that (for some) working on and successfully solving a math problem can bring a sense of beauty and order to the universe. A correct solution can be experienced as listening to the eternal voice of nature, speaking in its native language.

Let me thank some of the people who have helped me. Dan Sterling, David Hammett and Mitch Cohen gave invaluable advice regarding the technical elements of the text. Carrie Regenstein, Ann Cohen, Beth Kochen and Nettie Forsheit provided feedback on the comments. Matt Kinzelberg helped in the book's design and in bringing it into the public eye. Finally, my wife, Arleen, was there throughout with review, support and encouragement.

Good luck and enjoy.

Lewis Forsheit
Los Angeles
lew_forsheit@earthlink.net
March 2004

Part I

Problems

1A Golden Ratio

Given that a point on a line segment divides that segment into a larger portion and a smaller portion, derive the position of that point so that the ratio of the larger portion to the smaller portion is equal to the ratio of the entire segment to the larger portion.

1B Fibonacci Series

Given that a Fibonacci Series is one where each term is the sum of the previous two terms (and the first two terms have been specified), prove that the ratio of any term to its preceding term increasingly approximates the Golden Ratio as the series progresses.

2A   Pythagorean Theorem - Classic

Derive the classic proof of the Pythagorean Theorem: $a^2 + b^2 = c^2$
using the properties of similar triangles formed by a right triangle's altitude.

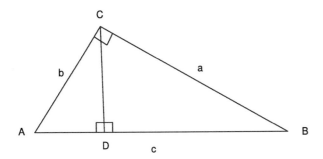

2B Pythagorean Theorem – Garfield's Trapezoid

Derive the proof of the Pythagorean Theorem: $a^2 + b^2 = c^2$
using the area properties of three triangles within a trapezoid
(attributed to President James Garfield.).

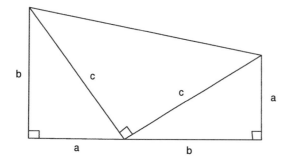

2C Pythagorean Theorem - Application to Inscribed Squares

Find the ratio of the area of a square inscribed in a semicircle, of radius r, to the area of a square inscribed in the whole circle of the same radius.

3A Law of Sines

Prove the Law of Sines for a triangle; namely: $\dfrac{a}{\sin A} = \dfrac{b}{\sin B} = \dfrac{c}{\sin C}$

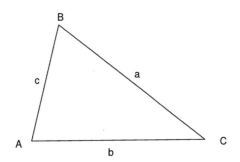

3B Law of Sines "Extended"

Prove the "Extended" Law of Sines for a triangle, namely:

$$\frac{a}{\sin A} = \frac{b}{\sin B} = \frac{c}{\sin C} = 2R$$

where R is the radius of the circumscribed circle.

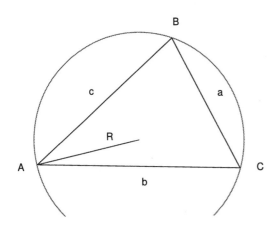

4A   Law of Cosines - Standard Configuration

Derive the Law of Cosines: $c^2 = a^2 + b^2 - 2ab\cos C$
using the standard proof diagram shown below.

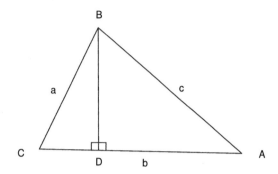

4B   Law of Cosines - Non-Standard Configuration

Derive the Law of Cosines: $c^2 = a^2 + b^2 - 2ab\cos C$
using the non-standard proof diagram shown below.

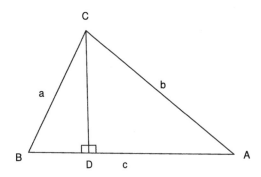

4C Law of Cosines - Application to Angle Bisectors

Using the Law of Cosines, show that $d^2 + mn = ab$ where d is the angle bisector that divides side c into segments m and n.

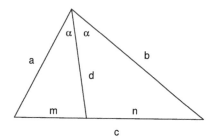

4D Law of Cosines - A Trigonometric Identity

Show that for three angles A, B and C where A + B + C = π rad

$$K = \frac{\sin A \cos A + \sin B \cos B + \sin C \cos C}{\sin A \sin B \sin C} = 2$$

4E   Law of Cosines - Diagonals of a Trapezoid

Find the diagonals, d_1 and d_2, of a trapezoid with
bases, b_1 and b_2, and sides, s_1 and s_2.

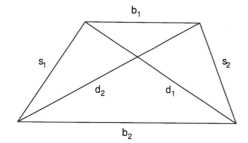

5A   Area of a Triangle – Three Side Formula

Prove that the area of a triangle, A, with sides a, b and c, is given by

$$A = \sqrt{s(s-a)(s-b)(s-c)} \quad \text{where s is the semi-perimeter, } \quad s = \frac{a+b+c}{2}.$$

5B Area of a Triangle - Inscribed Circle

Show that the radius, r, of a triangle's inscribed circle is:

$$r = \frac{\sqrt{s(s-a)(s-b)(s-c)}}{s} \quad \text{where s is the semi-perimeter} \quad s = \frac{a+b+c}{2}.$$

5C Area of a Triangle - Circumscribed Circle

Prove that the radius, R, of a triangle's circumscribed circle is given by:

$$R = \frac{abc}{4\sqrt{s(s-a)(s-b)(s-c)}} \quad \text{where s is the semi-perimeter} \quad s = \frac{a+b+c}{2}.$$

5D Area of a Triangle - One Side Formula

Derive a formula for the area of a triangle based on the length of one side and its adjacent angles.

6A Triangle Cerians

For the diagram shown, where d is called a "cerian," prove Stewart's Formula:

$$cd^2 + mnc = na^2 + mb^2$$

How does this formula change if d is the altitude? the median? the angle bisector?

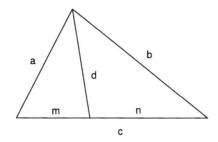

7A   Triangle with Equal Altitudes

Prove that if a triangle has two altitudes of equal length, it is an isosceles triangle.

7B   Triangle with Equal Medians

Prove that if a triangle has two medians of equal length, it is an isosceles triangle.

7C Triangles with Equal Angle Bisectors

Prove that if a triangle has two angle bisectors of equal length, it is an isosceles triangle.

8A Inscribed Quadrilateral – Specific Case

Find the radius of a semicircle whose inscribed quadrilateral has side lengths of 1, 2 and 3 and the fourth side is a diameter.

8B Inscribed Quadrilateral - General Case

Find the radius of a semicircle whose inscribed quadrilateral has side lengths of a, b and c and the fourth side is a diameter.

9A Factoring a Quartic Binomial – Specific Case

Factor: $x^4 + 4$.

9B Factoring a Quartic Binomial – General Case

Find as many methods as possible to factor: $x^4 + y^4$.

10A Factoring a Three Variable Quartic Polynomial

Factor: $2a^2b^2 + 2a^2c^2 + 2b^2c^2 - a^4 - b^4 - c^4$

11A   Square Rooting - Complex Numbers

Find the square roots of complex number a + bi.

11B Square Rooting - Irrational Numbers

Find the square root of a mixed irrational number $a + \sqrt{b}$.

12A   The Irrationality of the Square Root of Two

Prove that $\sqrt{2}$ is irrational.

That is, show that $\sqrt{2}$ is not equal to the quotient of two integers.

13A   A Quartic Equation – Specific Case

Solve for x: $x^4 + 5x^3 + 6x^2 + 5x + 1 = 0$.

13B   A Quartic Equation – General Case

Solve for x in the general case of the special quartic equation:

$x^4 + ax^3 + bx^2 + ax + 1 = 0.$

14A   Applying Imaginary Numbers

For the electric circuit shown, derive the amplitude, a, and phase, ϕ, of the output signal as a function of input voltage frequency, ω. Also find ω_0, the frequency that maximizes the amplitude, and ω_1 and ω_2, the frequencies at which the amplitude is $\frac{\sqrt{2}}{2}$ of its maximum. Lastly, find the bandwidth, $\omega_2 - \omega_1$.

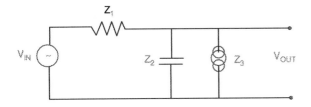

Given:

$$V_{IN} = \sin \omega t$$

$$V_{OUT} = a \sin(\omega t + \phi)$$

$$a = \frac{V_{OUT}}{V_{IN}} = \left([\text{Re}\,\frac{Z_P}{Z_1 + Z_P}]^2 + [\text{Im}\,\frac{Z_P}{Z_1 + Z_P}]^2 \right)^{\frac{1}{2}}$$

$$\phi = \tan^{-1}\left(\frac{\text{Im}\,\dfrac{Z_P}{Z_1 + Z_P}}{\text{Re}\,\dfrac{Z_P}{Z_1 + Z_P}} \right)$$

where

$$Z_1 = R$$

$$Z_2 = \frac{1}{j\omega C}$$

$$Z_3 = j\omega L$$

$Z_P = \dfrac{Z_2 Z_3}{Z_2 + Z_3}$ is the combined (parallel) impedance of Z_2 (the impedance of the capacitor) and Z_3 (the impedance of the inductor) connected in parallel and Z_1 is the impedance of the resistor.

Note: $j = \sqrt{-1} = i$ (which is the usual symbol for the imaginary number unit).

15A Imaginary Exponents

Using Euler's Formula: $e^{i\theta} = \cos\theta + i\sin\theta$, calculate the value of a^i and i^i, where a is a real number and $i = \sqrt{-1}$.

16A Probability

Our intrepid statistics student, Dewey Cheatham, was planning to take a test in his literature class and he had heard a rumor that it would be 20 True/False questions. Because he had not done the required reading, and had no clue at all what the book was about, his plan was to take random guesses for each answer. Dewey figured that, on average, he should get 50% correct and that his chances of getting a passing grade of at least 60% were pretty good. Unfortunately, when he actually took the exam, he saw that there were, in fact, 20 multiple choice questions, each with five possible answers. Needless to say, Dewey realized that by choosing his answers randomly, his chances for getting a passing grade went down considerably.

Using random guessing, what were Dewey's chances of passing with at least a 60% grade in the case of 20 True/False questions? What were they in the case of 20 five option multiple choice questions?

17A Confidence

Dewey has also just taken a nationwide exam in basic math skills along with 100,000 other students. On this exam, the questions are graded so that any score at all is possible between 0% and 100%. Here, Dewey was more prepared and received a grade of 70%. Dewey, figuring that he is really an average kind of guy, sitting atop the bell curve, believes that the national average is also 70%.

What is the required sample size of a survey, whose average score is 70%, to assure Dewey, to within 95% confidence level, that the national average is between 67% and 73%?

How would the required sample size change if Dewey had had a score of 80% and sought 95% confidence that the range was 77% to 83%?   If his score were 90% (seeking 95% confidence in the 87%-93% range)?

18A Correlation

It is now the end of the semester and Dewey is reflecting on his math class achievements. He studies the grades he received on the exams that were given twice a month and wants to see if he can learn something about himself. He would like to explore whether there were any outside (of math class) influences that affected his performance on the exams.

After pondering, Dewey concludes that there are three things that worry him on a daily basis: dealing with cold weather, running out of money and keeping up a good relationship with his girlfriend who is at another school. Could any of these factors be influencing his performance on math exams?

To see if any of these are possible, Dewey does some research from the weather service web site, his bank records and his telephone bills. He compiles the following table.

Date of Math Exam	Test Score on Math Exam	Temperature at 8:00 am	Account Balance	Phone Minutes Previous Night
Feb 1	92	22	450	28
Feb 15	85	19	475	25
Mar 1	98	21	275	32
Mar 15	81	29	325	20
Apr 1	86	41	375	30
Apr 15	88	38	300	29
May 1	90	50	400	31
May 15	95	52	410	35
Jun 1	94	48	390	35
Jun 15	93	73	380	35

How well correlated are the temperature, his account balance and his phone time, individually, with his math exam scores?

Are there any improvements that Dewey can make next year based on this information?

19A Arc Length

What is the length of the curve: $y = x^{3/2}$ from $x = 0$ to $x = a$?

20A Surface and Volume of a Solid of Revolution

What is the surface area and volume of the solid defined by the hyperbola $y = \dfrac{1}{x}$, for $0 < x \leq 1$, revolved around the y axis?

Can this three dimensional figure be painted?

21A Two Intersecting Cylinders

What is the volume of intersection between two right cylinders, both of radius r, whose axes intersect perpendicularly?

22A Integration

Find $\int (e^{ax} \cos bx)\,dx$

23A   Related Rates

A boy is standing b feet from the railroad track, and follows an oncoming train, traveling at v feet/sec, by turning his head. How fast will his head be turning when the train passes directly in front of him?

Part II

Solutions, Applications and Comments

Every effort has been made in the Problems statements and in the Solutions descriptions to present mathematics in its "pure" form. That is, no extraneous ideas have been added. The strength of this workbook ultimately rests on these problems and solutions, because it is through working on and studying them that students accrue the greatest benefit.

The applications and comments are designed for editorializing. Included are: additional notes on the math itself, descriptions of how the math is used in applications, pedagogy, some historical figures and their roles, personal experiences and notes, a little humor, etc.

It is recommended that students both work the problems and also discuss their respective applications and comments. Due to the many additional ways of addressing the problems themselves, there are an even far greater number of perspectives that can be contributed to the discussion. Students who have not worked the problems may also benefit from some of the insights in this section.

1A   Golden Ratio

Given that a point on a line segment divides that segment into a larger portion and a smaller portion, derive the position of that point so that the ratio of the larger portion to the smaller portion is equal to the ratio of the entire segment to the larger portion.

Solution

$$\underset{\phi \qquad\qquad\qquad 1}{\rule{300pt}{0.5pt}}$$

Let the length of the smaller portion of the line segment be defined as 1 and the larger portion be defined as ϕ. Then the Golden Ratio is $\phi/1 = \phi$.

Since the ratio of the larger portion to the smaller portion is the same as the ratio of the entire segment to the larger portion, we can set up the following equation:

$$\frac{\phi}{1} = \frac{\phi+1}{\phi}$$

This yields a quadratic equation and solution:

$$\phi^2 - \phi - 1 = 0$$

$$\phi = \frac{1+\sqrt{5}}{2} \approx 1.618...$$

We select the positive root for this geometric application.*

* If we select the negative root
$$\phi = \frac{1-\sqrt{5}}{2} \approx -0.618...$$

the interpretation is that the smaller portion has become the entire segment and that the ratio of the two portions are

$$\frac{\frac{\sqrt{5}-1}{2}}{1-\frac{\sqrt{5}-1}{2}} = \frac{1+\sqrt{5}}{2} \quad\text{or}\quad \frac{0.618...}{0.381...} \approx 1.618...$$

The Golden Ratio is one of the most amazing, yet relatively unknown, math concepts that is easily accessible by an intermediate algebra student. The geometric ratio, ϕ, formulation is straightforward and the resulting equation is a quadratic appropriate for students to solve when learning the Quadratic Formula.

54

The answer, ϕ, has a number of fascinating qualities, such as $\phi = 1.618...$ and $1/\phi = 0.618...$ (More on this later.)

Perhaps the most remarkable aspect of the Golden Ratio is that it is part of our every day lives and the public (in addition to math students) are totally unaware of it. From the time of the ancient Greeks, this ratio of 1.618 has been considered "perfect" and has been incorporated into countless artifacts as well as being found in nature.

I assigned a project on the Golden Ratio to my students, with each one required to find a different example. They found many, such as: the dimensions of the Parthenon; Da Vinci's Annunciation, St. Jerome and other drawings; Michaelangelo's Sistine Chapel, whose canvas dimensions, as well as the placement of the subjects is just this ratio. Dali's Sacrament of the Last Supper uses the ratio in several ways. Mozart, in piano sonatas, and Beethoven, in the Fifth Symphony, placed their signature themes at the Golden Ratio measures of the score. The students found the ratio in the Egyptian pyramids, the body dimensions of tigers and other animals, and plenty more, so the total list became very long.

Modern day advertisers have found that the shape of a rectangle in Golden Ratio proportions is considered the most pleasing to viewers so signs are often made that way. Many products such as cereal boxes also have Golden Ratio proportions. Even modern day credit cards have a length to width ratio of about 1.6 (which makes for an interesting double entendre of the Golden Ratio.)

The human body has many features that are in Golden Ratio proportions both in the limbs and on the face. Documentaries have been made based on the premise that the more facial features that are in Golden Ratio proportion, the more beautiful the face (both male and female). Commonly measured are the ratios of the width of the closed mouth to the width of the nose across the nostrils; the ratio of the vertical distance from pupils to lips and pupils to nose peak, and many others.

And, back at the math ranch, there are plenty more Golden Ratios to be found in geometric figures such as 18 degree triangles, pentagons, pentagrams, decagons, dodecagons, Penrose Tiles, rectangular and triangular equiangular spirals, etc.

Note that another interesting expression for ϕ comes from rewriting its equation as $\phi = \sqrt{1+\phi}$. Substituting for the ϕ under the radical, the value of $\phi = \sqrt{1+\phi}$, and continuing to do so, we get a continuing radical: $\phi = \sqrt{1+\sqrt{1+\sqrt{1+\sqrt{1+...}}}} \approx 1.618...$

The Golden Ratio is a fascinating topic that is worth spending time on in class as well as for assigning independent study. (See also Problem 14A.)

1B Fibonacci Series

Given that a Fibonacci Series is one where each term is the sum of the previous two terms (and the first two terms have been specified), prove that the ratio of any term to its preceding term increasingly approximates the Golden Ratio as the series progresses.

Solution

From the definition of the Fibonacci Series,

$$F_i = F_{i-1} + F_{i-2}$$ where F_i represents the ith term in the series.

If we divide F_i by F_{i-1}, we get

$$\frac{F_i}{F_{i-1}} = 1 + \frac{F_{i-2}}{F_{i-1}}$$

We can substitute for F_{i-1} the sum of $F_{i-2} + F_{i-3}$ and divide the numerator and denominator by F_{i-2}. We get

$$\frac{F_i}{F_{i-1}} = 1 + \frac{F_{i-2}}{F_{i-2} + F_{i-3}} = 1 + \frac{1}{1 + \frac{F_{i-3}}{F_{i-2}}}$$

Continuing this process indefinitely, as i becomes larger, yields the recurring fraction:

$$\frac{F_i}{F_{i-1}} = \cfrac{1}{1 + \cfrac{1}{1 + \cfrac{1}{1 + \cfrac{1}{1 + \ldots}}}}$$

Now we can compare this ratio with the Golden Ratio, ϕ.
We can write the equation for ϕ as:

$$\phi = 1 + \frac{1}{\phi}$$

If we replace the ϕ in the denominator with $\phi = 1 + \frac{1}{\phi}$, we get

$$\phi = 1 + \cfrac{1}{1 + \cfrac{1}{\phi}}$$

56

Repeating this process repeatedly, yields

$$\phi = \cfrac{1}{1+\cfrac{1}{1+\cfrac{1}{1+\cfrac{1}{1+...}}}}$$

This is the same recurring fraction as in the Fibonacci series.

Fibonacci Series were named for an Italian mathematician whose original name was Leonardo da Pisa and who explored many different aspects of algebra and geometry. The Series is what he is most famous for although his work in this area was not extensive.

Usually, math students are more familiar with Fibonacci Series than they are with the Golden Ratio so this problem is a good way of combining the two and extending the Golden Ratio study even further.

Additional work can be done comparing the Fibonacci ratio with the Golden Ratio. As i increases, each successive ratio gets closer to ϕ with the difference alternating between being positive and negative.

Fibonacci Series are also bountiful in nature. Petal and seed formations of many flowers, including daisy, columbine, sunflower, passion flower, chrysanthemum, etc. follow the Series pattern. Pine cones, pineapples and sea shells are also designed around this principle. If adjacent squares are constructed, starting with two squares of side length one, bordering on a square with side two, then three, five, eight, etc, a quarter circle can be drawn through each set of vertices, in succession, thereby forming a spiral in the exact shape of a human ear. The Fibonacci Series is also used to model the reproductive rate of rabbits.

2A Pythagorean Theorem - Classic

Derive the classic proof of the Pythagorean Theorem: $a^2 + b^2 = c^2$
using the properties of similar triangles formed by the altitude to the hypotenuse.

Solution

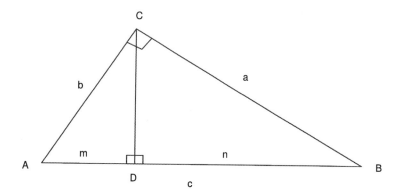

As shown in the diagram

CD is an altitude to side AB and

AC = b, CB = a, AB = c, AD = m and DB = n   (where c = m + n).

$\angle ACD = \angle CBD$ because they are both complementary to $\angle CAD$.

Thus $\triangle ACD \sim \triangle CDB \sim \triangle ABC$ because they have two equal angles.

The sides of similar triangles are in proportion, so

$$\frac{b}{m} = \frac{c}{b} \quad \text{or} \quad b^2 = cm \quad \text{and}$$

$$\frac{a}{n} = \frac{c}{a} \quad \text{or} \quad a^2 = cn \quad \text{Adding these equations gives}$$

$$a^2 + b^2 = cm + cn = c(m + n) \quad \text{or}$$

$$a^2 + b^2 = c^2$$

QED

This "classical" proof has an elegance in that it is a natural consequence of the study of similar triangles formed by the altitude of a right triangle. The Pythagorean Theorem is the workhorse of geometry and we will see it (along with its spin-off, the Law of Cosines) being used over and over again in solving the problems in this book.

One very important application of right angles is in the construction business. In the ancient world, ropes with knots were used to measure horizontally and vertically to assure perpendicularity. The hypotenuse is named for the Greek word for "subtending" which describes the expanse of the shorter legs. Today, the Pythagorean Theorem is no less important in construction; we are still quite dependent on it.

A fun experience for a geometry class is to ask the students to bring in a linen bed sheet to the next class. For that class, everyone wraps himself or herself in the sheet as if it were a Greek toga. Then, lead the class down the hall and through the building to examine the construction techniques, right angles, Pythagorean relationships, etc.

In addition to studying Greek geometry, two other Greek pedagogies are also in play: one is the Socratic method of asking questions; the other is the peripatetic style, which is walking around while talking and studying.

It is not really known if this is the proof that Pythagoras actually did himself. To that point, it is worth mentioning that Pythagoras had followers that attributed everything they did to him, so we don't really know. (The "Pythagorean Theorem" was clearly understood by the Babylonians a thousand years earlier, so we can only wonder what they called it then.) Pythagoras, it is told, killed a man who disclosed the secret proof that $\sqrt{2}$ is not a rational number. Still, the student is encouraged to try this proof, albeit at his or her own peril (see Problem 12A).

Finally, (literally), QED is an acronym for the Latin expression "Quod Erat Demonstrandum" which translates to "that which was to be demonstrated" or "that which was to be proven." For many years mathematicians used that expression to declare the successful end of a geometric proof. Modern students have, in the main, lost this practice, unfortunately, and, concomitantly, have lost an appropriate emphasis and elegance to the art form of geometric proofs. In a compromise gesture, one of my students has replaced the Latin in favor of the French "Voila" ("there it is").

2B Pythagorean Theorem – Garfield's Trapezoid

Derive the proof of the Pythagorean Theorem: $a^2 + b^2 = c^2$
using the area properties of three triangles within a trapezoid
(attributed to President James Garfield.).

Solution

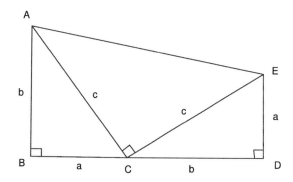

As shown in the diagram

$AB = b$, $BC = a$, $AC = c$, $CD = b$, $DE = a$, $CE = c$

The area of $\triangle ABC = \frac{1}{2}\, ab$

The area of $\triangle CDE = \frac{1}{2}\, ab$

The area of $\triangle ACE = \frac{1}{2}\, c^2$

The area of trapezoid ABDE is one half the altitude (a + b) times
the sum of the bases (a + b) or $\frac{1}{2}\,(a + b)^2$.

Added together, the areas of these three triangles equal the area of the trapezoid
ABDE.

$\frac{1}{2}\, ab + \frac{1}{2}\, ab + \frac{1}{2}\, c^2 = \frac{1}{2}\,(a + b)^2$. Multiplying out and collecting terms yields

$a^2 + b^2 = c^2$

QED

Certainly, the earliest presidents had classical educations and were well versed in Latin, Greek, mathematics, literature, etc. President Garfield also had an interest in such academic affairs and he developed this proof of the Pythagorean Theorem while he was a member of the U.S. House of Representatives. Apparently, the debate over legislative bills in the House did not require his full attention, which enabled Garfield to multi-task so successfully. There is no reason to believe that his creating this derivation was a contributing factor to the circumstances leading to his assassination.

The method used here, of adding three triangles' areas and equating the sum to the area of a trapezoid, will be used again in Solution 2 of Problem 4D.

2C Pythagorean Theorem - Application to Inscribed Squares

Find the ratio of the area of a square inscribed in a semicircle, of radius r, to the area of a square inscribed in the whole circle of the same radius.

Solution

The two squares are shown in the diagram below.

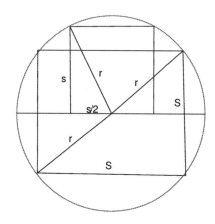

Applying the Pythagorean Theorem to the square inscribed in the semicircle:

$$s^2 + \left(\frac{s}{2}\right)^2 = r^2$$

where r is the radius of the semicircle

and s is the side of the small square.   Then,

$$\frac{5s^2}{4} = r^2$$

and

$$s^2 = \frac{4r^2}{5}$$

where s^2 is the area of the small square.

Applying the Pythagorean Theorem to the square inscribed in the circle:

$S^2 + S^2 = (2r)^2$ where S is the side of the large SQUARE
$2S^2 = 4r^2$ and
$S^2 = 2r^2$ where S^2 is the area of the large SQUARE

Then the ratio of the areas is:

$$\frac{A_{square}}{A_{SQUARE}} = \frac{\dfrac{4r^2}{5}}{2r^2} = \frac{2}{5}$$

This is a good learning problem for geometry students because it requires a little creativity in visualizing the unusual case of a square inscribed in a semicircle. Once drawn, the student should not have too much difficulty in establishing the equations for solution. This problem is actually a theorem of its own but is rarely seen in geometry texts.

Assuming students are familiar with the definition and concept of circumscribed circles, this problem is an excellent "advanced problem" while studying the Pythagorean Theorem. Other problems that study circumscribed circles are Problems 3B and 5C.

Another important lesson from this problem is the idea of parameterizing geometric relationships. The solution does not require calculating a specific value of the squares' areas but, instead, the idea is to relate everything to the length of the radius. This is an important method or approach in conceptualizing relative proportions, especially the idea that areas vary as the second order of length.

This problem also requires a modicum of algebraic facility in the reduction of multi-layered fractions. Several problems in this text require extensive algebraic manipulation to solve. The integration of geometry and algebra is usually kept to a minimum in most texts (but not here).

Another perspective on this problem is that it gives us some insight into the field of interior design. Placing shapes into existing spaces, finding relative sizes between objects in odd configurations and maximizing living and work areas are important commercial applications of this form of geometry. Computer programs now offer support to interior designers to help create both functional and aesthetic décor.

3A Law of Sines

Prove the Law of Sines for a triangle; namely: $\dfrac{a}{\sin A} = \dfrac{b}{\sin B} = \dfrac{c}{\sin C}$

Solution

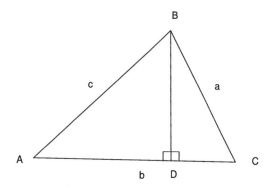

In the diagram shown, $BD \perp AC$ so

$BD = c\sin A$ and $BD = a\sin C$

Therefore, $c\sin A = a\sin C$ or

$$\frac{c}{\sin C} = \frac{a}{\sin A}$$

Similarly, by drawing an altitude from C to side AB, we can show that

$$\frac{b}{\sin B} = \frac{a}{\sin A}$$

Therefore, $\quad \dfrac{a}{\sin A} = \dfrac{b}{\sin B} = \dfrac{c}{\sin C}$

QED

Learning the Law of Sines requires a basic knowledge of trigonometric functions. Perhaps the most famous mnemonic device is the term SOHCAHTOA that helps students remember that sine is opposite over hypotenuse, cosine is adjacent over hypotenuse and tangent is opposite over adjacent.

Pedagogies have been developed for the exploration of and the learning about trigonometric functions. One that I have used successfully is for students to research physics topics that have sine wave functions.

For example, the student can research how: an electric generating station creates alternating current in the form of a sine wave, why a violin string vibrates and creates a note of music, how to design a variable LC circuit in a radio to receive different stations, and how to generate different kinds of sine waves to enable both AM and FM radio signals. The student explores the situation, finds the equations that describe the physics (often well beyond their current level of understanding in math or physics), identifies the sine wave solutions that evolve from the equations, and then they are on their own in the fascinating journey of how it all gets enacted and applied. The sine wave parameters of amplitude, frequency, phase and offset become universals both in nature and in invention.

The Law of Sines, itself, has many practical applications. One is that aircraft guidance systems compute how large an angle, α, is needed to head into the wind so that the ultimate direction that the plane flies will be on course.

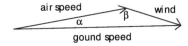

Another application of the Law of Sines is Snell's Law regarding the refraction of light through glass. This is used in the design of optical lenses and prisms.

3B   Law of Sines "Extended"

Prove the "Extended" Law of Sines for a triangle, namely:

$$\frac{a}{\sin A} = \frac{b}{\sin B} = \frac{c}{\sin C} = 2R$$

where R is the radius of the circumscribed circle.

Solution

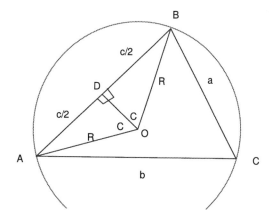

In the diagram, $\angle BOA$ is a central angle and $\angle BCA$ is an inscribed angle that both subtend arc BA.

Therefore, $\angle BOA = 2\angle BCA = 2\angle C$

Construct $OD \perp AB$ so $\triangle BOD \cong \triangle DOA$ (SSS) and OD bisects $\angle BOA$.

Then, $\angle BOD = \angle DOA = \angle C$ as shown in the diagram.

Looking at $\triangle DOB$, $\sin C = \dfrac{c}{2R}$ where R is the radius of the circumscribed circle.

Similarly, relationships for $\angle A$ and $\angle B$ are:

$$\sin A = \frac{a}{2R} \qquad \sin B = \frac{b}{2R}. \quad \text{Then,}$$

$$\frac{a}{\sin A} = \frac{b}{\sin B} = \frac{c}{\sin C} = 2R$$

QED

There are few texts or classroom exercises that explore the Law of Sines in this extended manner. Students rarely get to see that the ratio of the Law of Sines actually has a meaning beyond the equivalency usually stated. Relating the ratio to the circumscribed circle brings a whole new dimension to the Law of Sines and also provides us with the ability to compare the ratio to those of other triangles, based on the size of the circumscribed circle.

4A Law of Cosines - Standard Configuration

Derive the Law of Cosines: $c^2 = a^2 + b^2 - 2ab\cos C$
using the standard proof diagram shown below.

Solution

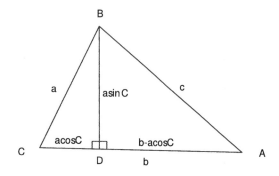

As shown in the diagram, with $BD \perp AC$, and

$AC = b$, $AB = c$, $BC = a$, then

$CD = a\cos C$, $BD = a\sin C$ and $AD = b - a\cos C$

Using the Pythagorean Theorem for $\triangle ADB$,

$c^2 = a^2\sin^2 C + (b - a\cos C)^2$

$c^2 = a^2\sin^2 C + b^2 - 2ab\cos C + a^2\cos^2 C$

Since $a^2\sin^2 C + a^2\cos^2 C = a^2(\sin^2 C + \cos^2 C) = a^2(1) = a^2$,

$c^2 = a^2 + b^2 - 2ab\cos C$

QED

There are many proofs of the Law of Cosines. This one is probably the most frequently used in trigonometry texts. It is included because it is a critical proof to know and because it introduces the subsequent problems.

The Law of Cosines is often taught as a "modification" or "adjustment" to the Pythagorean Theorem. The first terms $c^2 = a^2 + b^2$ are the same. The term $-2ab\cos C$ is an "adjustment" made to the equation to compensate for angle C not being 90 degrees.

It is worth exploring what happens to the equation as C varies:

If $C = 0$ deg, $\cos C = +1$, then $c^2 = a^2 - 2ab + b^2$ and $c = a - b$.
If $C = 180$ deg, $\cos C = -1$, then $c^2 = a^2 + 2ab + b^2$ and $c = a + b$.
If $C = 90$ deg, $\cos C = 0$, then $c^2 = a^2 + b^2$ the Pythagorean Theorem.
If $C = 60$ deg, $\cos C = \frac{1}{2}$, then $c^2 = a^2 - ab + b^2$ and if $a = b$, then $c = a = b$, as expected.
If $C = 120$ deg, $\cos C = -\frac{1}{2}$, then $c^2 = a^2 + ab + b^2$ and if $a = b$, then $c = a\sqrt{3} = b\sqrt{3}$.

As is the case of the Law of Sines, the Law of Cosines gets plenty of work-out in the "real world." As shown in the diagram in the Problem 3A commentary, the same guidance system that uses the Law of Sines to compute the required angle for heading into the wind, also uses angle β and the Law of Cosines for computing the aircraft ground speed.

The Law of Cosines is also used extensively in surveying, to measure distances in areas that are not easily accessible.

4B Law of Cosines - Non-Standard Configuration

Derive the Law of Cosines: $c^2 = a^2 + b^2 - 2ab\cos C$
using the non-standard proof diagram shown below.

Solution

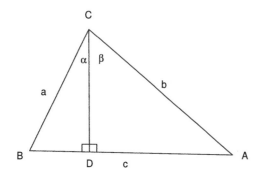

As shown in the diagram with: $CD \perp AB$,

$AC = b$, $CB = a$, $BA = c$, $\angle C = \angle\alpha + \angle\beta$

$BD = a\sin\alpha$ and $DA = b\sin\beta$. Then

$AB = c = a\sin\alpha + b\sin\beta$. Squaring the equation yields

$c^2 = a^2\sin^2\alpha + b^2\sin^2\beta + 2ab\sin\alpha\sin\beta$

Replacing $a^2\sin^2\alpha$ and $b^2\sin^2\beta$ with $a^2(1 - \cos^2\alpha)$ and $b^2(1 - \cos^2\beta)$, respectively, gives

$c^2 = a^2 + b^2 - a^2\cos^2\alpha - b^2\cos^2\beta + 2ab\sin\alpha\sin\beta$

But $CD = a\cos\alpha = b\cos\beta$

and $a^2\cos^2\alpha = b^2\cos^2\beta = ab\cos\alpha\cos\beta$. Combining, yields

$c^2 = a^2 + b^2 - 2ab\cos\alpha\cos\beta + 2ab\sin\alpha\sin\beta$ or

$c^2 = a^2 + b^2 - 2ab(\cos\alpha\cos\beta - \sin\alpha\sin\beta)$.

We recognize the last term in parenthesis as the cosine of the sum of two angles:

$\cos\alpha\cos\beta - \sin\alpha\sin\beta = \cos(\alpha + \beta) = \cos C$. Then

$c^2 = a^2 + b^2 - 2ab\cos C$

QED

Sometimes, a student will inadvertently try to prove the Law of Cosines as shown here. This is especially more difficult because the study of trigonometric functions for the sum of two angles is taught well after the Law of Cosines.

Usually, the classical mathematics perspective of any derivation is to use the simplest construction. This more difficult approach shows that the proof can be done even if the simplest construction is not found.

4C   Law of Cosines – Application to Angle Bisectors

Using the Law of Cosines, show that $d^2 + mn = ab$ where d is the angle bisector that divides side c into segments m and n.

Solution

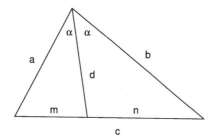

Using the Law of Cosines,

$m^2 = a^2 + d^2 - 2ad\cos\alpha$

$n^2 = b^2 + d^2 - 2bd\cos\alpha.$

Solving each for $\cos\alpha$ and setting the results equal to each other:

$$\frac{m^2 - a^2 - d^2}{-2ad} = \frac{n^2 - b^2 - d^2}{-2bd}$$

Multiplying out and collecting terms yields:

$d^2(a - b) + (bm^2 - an^2) = ab(a - b)$

When d is an angle bisector:

$$\frac{m}{a} = \frac{n}{b} \quad \text{or} \quad \frac{m}{n} = \frac{a}{b} \quad \text{or} \quad \frac{n}{m} = \frac{b}{a}. \ *$$

Substituting, $bm^2 - an^2 = mn\left(\dfrac{bm}{n} - \dfrac{an}{m}\right) = mn\left(b \cdot \dfrac{a}{b} - a \cdot \dfrac{b}{a}\right) = mn(a - b)$ above, yields

$d^2(a - b) + mn(a - b) = ab(a - b)$ or

$d^2 + mn = ab$

QED

* To prove this assertion, see in the diagram below,
  $\triangle ABC$, $BC = a$, $AC = b$, $AB = c$,

CD is an angle bisector with $\angle BCD = \angle ACD$, and $BD = m$ and $DA = n$.

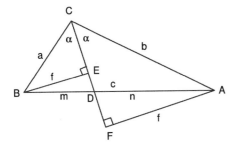

Construct a perpendicular to CD from B with an intersection at E, with $BE = f$.
Construct a perpendicular to CD from A with an intersection at F, with $AF = g$.

Since $\angle BED = \angle AFD$ (both right angles) and $\angle BDE = \angle ADF$ (vertical angles),
$\triangle BED \sim \triangle ADF$.

Then, $m/n = f/g$ since corresponding sides of similar triangles are proportional.

Since $\angle BED = \angle AFD$ (both right angles) and $\angle BCE = \angle ACF$ (given),
$\triangle BEC \sim \triangle ACF$.

Then, $f/g = a/b$ since corresponding sides of similar triangles are proportional
and $m/n = a/b$ because equals equal to the same thing are equal.

                                                            QED

We will see this equation again in Problem 6A. The equations are the same because, ultimately, each answer is derived through implementation of the Pythagorean Theorem.

4D Law of Cosines - A Trigonometric Identity

Show that for three angles, A, B and C, where $A + B + C = \pi$ rad

$$K = \frac{\sin A \cos A + \sin B \cos B + \sin C \cos C}{\sin A \sin B \sin C} = 2$$

Solution 1

Since $C = \pi - (A + B)$,

$\sin C = \sin[\pi - (A + B)] = \sin(A + B) = \sin A \cos B + \cos A \sin B$ and

$\cos C = \cos[\pi - (A+B)] = -\cos(A + B) = \sin A \sin B - \cos A \cos B$. Then

$\sin C \cos C = \sin^2 A \sin B \cos B - \cos^2 B \sin A \cos A + \sin^2 B \sin A \cos A - \cos^2 A \sin B \cos B$ or

$\sin C \cos C = \sin A \cos A(\sin^2 B - \cos^2 B) + \sin B \cos B(\sin^2 A - \cos^2 A)$.   Thus,

$\sin A \cos A + \sin B \cos B + \sin C \cos C =$
$$\sin A \cos A(1 + \sin^2 B - \cos^2 B) + \sin B \cos B(1 + \sin^2 A - \cos^2 A)$$

$\sin A \cos A + \sin B \cos B + \sin C \cos C = \sin A \cos A(2\sin^2 B) + \sin B \cos B(2\sin^2 A)$. So

$$K = \frac{\sin A \cos A(2\sin^2 B) + \sin B \cos B(2\sin^2 A)}{\sin A \sin B(\sin A \cos B + \cos A \sin B)}$$

$$K = 2 \cdot \frac{\sin A \cos A(\sin^2 B) + \sin B \cos B(\sin^2 A)}{\sin^2 A \sin B \cos B + \sin^2 B \sin A \cos A} = 2$$

Solution 2

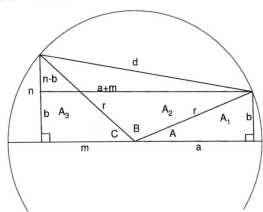

In the diagram, we see

$$\sin A = \frac{b}{r} \qquad \cos A = \frac{a}{r}$$

$$\sin C = \frac{n}{r} \qquad \cos C = \frac{m}{r}$$

To find sinB and cosB, let $A_p = A_1 + A_2 + A_3$

where A_p is the area of the trapezoid, and A_1, A_2, and A_3 are the areas of the three

triangles with angles A, B and C, respectively.

Using the formula for the area of a trapezoid as ½ the product of the altitude (a+m) and the sum of the bases (b+n),

½(a + m)(b + n) = ½(ab) + ½(r²)sinB + ½(mn)

Solving for sinB yields,

$$\sin B = \frac{mb + an}{r^2}$$

Using the Pythagorean Theorem, we have

d² = (a + m)² + (n − b)²

And, using the Law of Cosines yields:

$$d^2 = r^2 + r^2 - 2r^2\cos B$$

Setting both expressions for d^2 equal to each other and solving for $\cos B$ yields

$$\cos B = \frac{bn - am}{r^2}$$

Substituting in the formula for K results in:

$$K = \frac{\dfrac{ab}{r^2} + \left(\dfrac{mb + an}{r^2}\right)\left(\dfrac{bn - am}{r^2}\right) + \dfrac{mn}{r^2}}{\dfrac{bn}{r^2}\left(\dfrac{mb + an}{r^2}\right)}$$

Multiplying by r^4 and collecting like terms, yields

$$K = \frac{r^2(ab) + r^2(mn) + ab(n^2 - m^2) + mn(b^2 - a^2)}{bn(mb + an)}$$

Substituting $m^2 + n^2$ for the r^2 coefficient of ab and $a^2 + b^2$ for the r^2 coefficient of mn,

$$K = \frac{(m^2 + n^2)(ab) + (a^2 + b^2)(mn) + ab(n^2 - m^2) + mn(b^2 - a^2)}{bn(mb + an)}$$

Then multiplying out and collecting like terms yields

$$K = \frac{2abn^2 + 2mnb^2}{bn(mb + an)} = \frac{2an + 2mb}{mb + an} = 2$$

QED

My research has not found this identity in any text, website or coursework, so it may be original here. Or, more likely, it may be documented in sources with which I am just not yet familiar.

Since the angles of a triangle add to π rad, this identity applies to all triangles.

Then, a more compact form of this identity could read:

For the three angles A, B and C of a triangle,

$$\frac{\sin 2A + \sin 2B + \sin 2C}{\sin A \sin B \sin C} = 4$$

This "new" formula may seem complex, but it adds an increased understanding of a triangle's symmetry that is not obvious to the eye.

Special case check: If A is a right angle, sin2A = 0 and sinA = 1. Substituting, yields

$$\frac{\sin 2B + \sin 2C}{\sin B \sin C} = 4$$

Since B + C = 90 degs, sinC = cosB, and
since 2B + 2C = 180 degs, sin2C = sin2B.

Then,

$$\frac{2 \sin 2B}{\sin B \cos B} = \frac{4 \sin B \cos B}{\sin B \cos B} = 4$$

which is an identity, and confirms the special case.

Solution 1 is the classical method of proving a trigonometric identity. That is, substituting equivalent expressions for the various terms in the expression. Solution 2 is a geometric proof that gives a clearer view of how the angles interact with each other.

As a closing comment, note that, in Solution 2, setting up the area of the trapezoid as the sum of the areas of the triangles and then deriving useful relationship information, is exactly the same process as in the Garfield Proof of the Pythagorean Theorem (see Problem 2B).

4E Law of Cosines - Diagonals of a Trapezoid

Find the diagonals, d_1 and d_2, of a trapezoid with
bases, b_1 and b_2, and sides, s_1 and s_2.

Solution

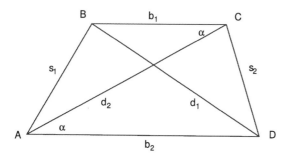

As seen in the diagram, in trapezoid ABCD, $\angle ACB = \angle CAD = \alpha$ since $AD \parallel BC$.

Using the Law of Cosines:

$$\cos \alpha = \frac{b_1^2 + d_2^2 - s_1^2}{2b_1 d_2} = \frac{b_2^2 + d_2^2 - s_2^2}{2b_2 d_2}$$

Canceling $2d_2$ from both sides of the equation, then multiplying out and collecting like terms yields

$$d_2^2 (b_2 - b_1) = b_2 (b_1 b_2 + s_1^2) - b_1 (b_1 b_2 + s_2^2)$$

Then,

$$d_2 = \sqrt{\frac{b_2 (b_1 b_2 + s_1^2) - b_1 (b_1 b_2 + s_2^2)}{b_2 - b_1}}$$

and, similarly,

$$d_1 = \sqrt{\frac{b_2 (b_1 b_2 + s_2^2) - b_1 (b_1 b_2 + s_1^2)}{b_2 - b_1}}$$

QED

For the case where ABCD is an isosceles trapezoid, the solutions for the diagonals

reduce to $d = \sqrt{s^2 + b_1 b_2}$

where $d = d_1 = d_2$ and $s = s_1 = s_2$.

This result can also be obtained by considering that

$$d \cos \alpha = \frac{b_2 - b_1}{2} + b_1 \qquad \text{Then,}$$

$$\cos \alpha = \frac{b_2 + b_1}{2d}$$

Using the Law of Cosines and substituting for $\cos \alpha$ gives:

$$s^2 = b_1^2 + d^2 - 2b_1 d \frac{(b_2 + b_1)}{2d} \qquad \text{and}$$

$$d = \sqrt{s^2 + b_1 b_2} \quad \text{as before.}$$

5A Area of a Triangle – Three Side Formula

Prove that the area of a triangle, A, with sides a, b and c, is given by

$A = \sqrt{s(s-a)(s-b)(s-c)}$ where s is the semi-perimeter, $s = \dfrac{a+b+c}{2}$.

Solution

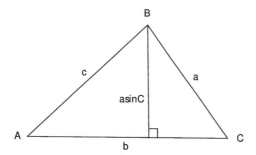

The area of the triangle can be written as $A = \frac{1}{2}(ab)\sin C$

Using the trigonometric identity $\sin C = \sqrt{1-\cos^2 C}$

and the Law of Cosines, $\cos C = \dfrac{c^2 - a^2 - b^2}{-2ab}$, we get

$$A = \frac{ab}{2}\sqrt{1-\left(\frac{c^2-a^2-b^2}{-2ab}\right)^2} = \frac{ab}{2}\cdot\frac{1}{2ab}\cdot\sqrt{4a^2b^2 - (c^2-a^2-b^2)^2}$$

Multiplying out the radical and canceling, yields

$$A = \frac{1}{4}\cdot\sqrt{2a^2b^2 + 2b^2c^2 + 2a^2c^2 - a^4 - b^4 - c^4}$$

Factoring

$2a^2b^2 + 2b^2c^2 + 2a^2c^2 - a^4 - b^4 - c^4 = (a + b + c)(-a + b + c)(a - b + c)(a + b - c)$
yields

$$A = \sqrt{\left(\frac{a+b+c}{2}\right)\left(\frac{-a+b+c}{2}\right)\left(\frac{a-b+c}{2}\right)\left(\frac{a+b-c}{2}\right)} \quad \text{or}$$

$A = \sqrt{s(s-a)(s-b)(s-c)}$

QED

The well known common formula for the area of a triangle is A = ½ bh where b is the base and h is the height, or altitude to side b. Since b = asinC, the next most common formula for the area of a triangle is A = ½ absinC.

Considering the usefulness of being able to calculate the area of a triangle knowing only the lengths of the sides, the three side area formula is rarely taught, little appreciated or even mentioned.

The derivation shown has a rare combination of, or interaction between, sines and cosines. And, the form of the equation may also be unique in that it may be the only one that uses the variable s, the semi-perimeter. This is done because it makes the formula more compact and easier to use.

Also, take particular note of the factoring step in the derivation:
$2a^2b^2 + 2b^2c^2 + 2a^2c^2 - a^4 - b^4 - c^4 = (a + b + c)(- a + b + c)(a - b + c)(a + b - c)$.
This factoring is not immediately or intuitively obvious and has been included later as Problem 10A.

All of the above is even more remarkable in that this formula for the area of a triangle is attributed to the ancient Greek, Heron of Alexandria.

Since owning land was always an important aspect of civilization, accurate measurements of land area were highly valued. Measuring the area of a quadrilateral can be inaccurate due to the probable non-perpendicularity of the sides. Measuring the area of a triangle, using only the value of the sides – and no angles, became a far more accurate measure. Any quadrilateral can be divided into two triangles by drawing a diagonal, so Heron's formula was widely used.

As an extended study of this topic, consider the area of a trapezoid with sides, s_1 and s_2, and bases, b_1 and b_2. Using the two triangle method, the area would be:

$$A = \frac{1}{4}\sqrt{(s_1 + b_1 + d_2)(-s_1 + b_1 + d_2)(s_1 - b_1 + d_2)(s_1 + b_1 - d_2)} + \frac{1}{4}\sqrt{(s_2 + b_2 + d_2)(-s_2 + b_2 + d_2)(s_2 - b_2 + d_2)(s_2 + b_2 - d_2)}$$

where the diagonal length, d_2, derived in Problem 4E, is:

$$d_2 = \sqrt{\frac{b_2(b_1b_2 + s_1^2) - b_1(b_1b_2 + s_2^2)}{b_2 - b_1}}$$

For even further study, the area of a trapezoid can be derived using the formula, $A = h(b_1 + b_2)/2$, with Pythagorean relationships to solve for h in terms of the sides and bases. The result is $A = \frac{b_2 + b_1}{4(b_2 - b_1)}\sqrt{2(s_1^2 + s_2^2)(b_2 - b_1)^2 - (s_1^2 - s_2^2)^2 - (b_2 - b_1)^4}$.

(Thanks go to my friends, John and Mary Calvin, for suggesting this problem to me.)

A truly yeoman task would be to show that the two formulas for the area of a trapezoid are equivalent.

5B   Area of a Triangle - Inscribed Circle

Show that the radius, r, of a triangle's inscribed circle is:

$$r = \frac{1}{s}\sqrt{s(s-a)(s-b)(s-c)} \quad \text{where s is the semi-perimeter} \quad s = \frac{a+b+c}{2}.$$

Solution

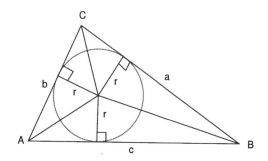

From the diagram

$$A_t = A_1 + A_2 + A_3$$

where A_t is the area of the entire triangle and A_1, A_2 and A_3 are the areas of the three smaller triangles formed by the triangle's sides and the lines from the vertices to the center of the inscribed circle.

Then,

$$A_t = \frac{1}{2}(ar) + \frac{1}{2}(br) + \frac{1}{2}(cr) = \frac{1}{2}(a+b+c)r = sr$$

and

$$r = \frac{A_t}{s} = \frac{1}{s}\sqrt{s(s-a)(s-b)(s-c)}$$

$$\text{QED}$$

This is a straightforward application of the three side Area Formula. Inscribed circles (as well as circumscribed circles) are often found in product designs. Since spheres are the geometric shape that maximizes the ratio of volume to surface area, many products have internal spheres for liquid storage.

5C   Area of a Triangle - Circumscribed Circle

Prove that the radius, R, of a triangle's circumscribed circle, is given by:

$$R = \frac{abc}{4\sqrt{s(s-a)(s-b)(s-c)}} \quad \text{where s is the semi-perimeter} \quad s = \frac{a+b+c}{2}.$$

Solution 1

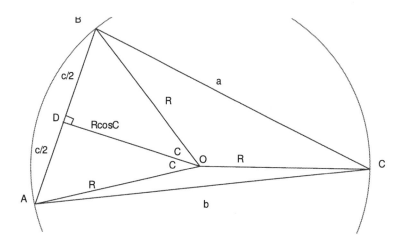

In the diagram, $\angle BOA$ is a central angle and $\angle BCA$ is an inscribed angle that both subtend arc BA.

Construct $OD \perp AB$. Then, $\triangle BOD \cong \triangle DOA$, OD bisects $\angle BOA$ and $\angle BOD = \angle DOA = \angle C$ (as in Problem 2B). Similar relationships exist for $\angle A$ and $\angle B$.

Then, $A_t = A_1 + A_2 + A_3$ where A_t is the area of the entire triangle and A_1, A_2 and A_3 are the three smaller triangles made by the triangle's sides and their respective radii of the circumscribed circle. The altitude of each small triangle is R times the cosine of the central angle. So,

$$A_t = \frac{1}{2}(aR\cos A + bR\cos B + cR\cos C)$$

Using the Law of Cosines for the large triangle,

$$\cos A = \frac{b^2 + c^2 - a^2}{2bc} \quad \text{and the counterpart formulas for cosB and cosC, we get}$$

$$A_t = \frac{R}{4}\left(a \cdot \frac{b^2+c^2-a^2}{bc} + b \cdot \frac{a^2+b^2-c^2}{ac} + c \cdot \frac{a^2+b^2-c^2}{ab} \right)$$

Multiplying out and combining fractions yields

$$A_t = \frac{R}{4abc}(2a^2b^2 + 2b^2c^2 + 2a^2c^2 - a^4 - b^4 - c^4)$$

From Problem 5A, we know that the value of the polynomial in the brackets, $2a^2b^2 + 2b^2c^2 + 2a^2c^2 - a^4 - b^4 - c^4$, equals $16A_t^2$. Then $A_t = \dfrac{R}{4abc}(16A_t^2)$

Solving for R gives:

$$R = \frac{abc}{4A_t} = \frac{abc}{4\sqrt{s(s-a)(s-b)(s-c)}}$$

QED

Solution 2

As in Solution 1,

$$A_t = \frac{1}{2}(aR\cos A + bR\cos B + cR\cos C)$$

Now, instead, we make the substitutions:
$a = 2R\sin A \quad b = 2R\sin B \quad c = 2R\sin C \qquad$ and get
$A_t = R^2(\sin A\cos A + \sin B\cos B + \sin C\cos C)$

Dividing by abc and substituting yields

$$\frac{A_t}{abc} = \frac{R^2(\sin A\cos A + \sin B\cos B + \sin C\cos C)}{8R^3\sin A\sin B\sin C}$$

Solving for R yields
$$R = \frac{abc}{8A_t}K \quad \text{where}$$

$$K = \frac{\sin A\cos A + \sin B\cos B + \sin C\cos C}{\sin A\sin B\sin C} = 2 \quad \text{(per the solution in Problem 4D).}$$

Then, $\quad R = \dfrac{abc}{4\sqrt{s(s-a)(s-b)(s-c)}}$

QED

Problems 5B and 5C provide quantitative answers to our intuitive appreciation that inscribed and circumscribed circles are dependent on the lengths of a triangle's sides. This idea can be seen as an extension of the commonly used "side-side-side" proof of triangles' congruency. That is, the three lengths create a polygon whose area, inscribed and circumscribed circles' radii are all uniquely determined.

Solution 2 refers to the solution in Problem 5A. In classical terms, Problem 5A is a "lemma" of Problem 5C. A lemma is a "subsidiary" or "helping" theorem used in proving a larger proposition. The old joke is to think of the lemma and the main theorem, together, as a "dilemma."

5D Area of a Triangle - One Side Formula

Derive a formula for the area of a triangle based on the length of one side and its adjacent angles.

Solution

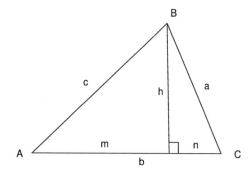

A = ½ bh.

As shown in the diagram,

h = mtanA = ntanC

and m + n = b.

Substituting,

(b − n)tanA = ntanC

btanA = n(tanA + tanC)

$$n = \frac{b \tan A}{\tan A + \tan C}$$

Then,

$$h = \frac{b \tan A \tan C}{\tan A + \tan C}$$

Area $A = \frac{1}{2}bh = \frac{1}{2}\frac{b^2 \tan A \tan C}{\tan A + \tan C}$ or, after diving by tanAtanC,

$$A = \frac{b^2}{2(\cot A + \cot C)}$$

QED

We began with a formula for the area of a triangle based on two sides and the included angle, and then proceeded to derive a formula using all three sides. Now, we are also able to do so with one side and its adjacent angles.

These formulas correspond to the three methods of proving uniqueness and congruency of triangles. They are: SAS (side-angle-side), SSS (side-side-side) and ASA (angle-side-angle), respectively.

We can now think in terms of four formulas for the area of a triangle

$$A = \frac{1}{2}\,bh = \frac{1}{2}\,ab\sin C = \sqrt{s(s-a)(s-b)(s-c)} = \frac{b^2}{2(\cot A + \cot C)}$$

Of course, the area of a triangle cannot be calculated on the basis of three angles because such a triangle in not unique. There can be an infinite number of triangles that are similar to it.

6A Triangle Cerians

For the diagram shown, where d is called a "cerian," prove Stewart's Formula:
$cd^2 + mnc = na^2 + mb^2$

How does this formula change if d is the altitude? the median?   the angle bisector?

Solution

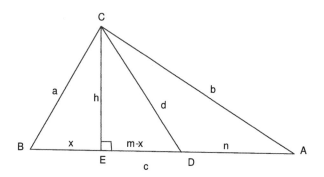

Construct an altitude from vertex C to a point, E, on side AB.

Let the length of that altitude be h, as shown in the diagram.

Also, let BE = x and ED = m − x.

Using the Pythagorean Theorem:

$x^2 + h^2 = a^2$

$(m - x)^2 + h^2 = d^2$

$(c - x)^2 + h^2 = b^2$ or $h^2 = b^2 - (c - x)^2$. So,

$x^2 + b^2 - (c - x)^2 = a^2$. Solving for x yields:

$$x = \frac{a^2 + c^2 - b^2}{2c}$$

Combining results gives:

$$\left(m - \frac{a^2 + c^2 - b^2}{2c}\right)^2 + b^2 - \left(c - \frac{a^2 + c^2 - b^2}{2c}\right)^2 = d^2$$

Now, multiplying the equation by 2c and replacing the $2c^2$ term by $2c(m+n)$, simplifies to:

$-mnc + (a^2 - b^2)n + cb^2 = cd^2$

Rearranging gives

$cd^2 + mnc = cb^2 + n(a^2 - b^2)$

$cd^2 + mnc = (m + n)b^2 + na^2 - nb^2$ and

$cd^2 + mnc = na^2 + mb^2$

QED

When d is an altitude,

$a^2 = m^2 + d^2$

$b^2 = n^2 + d^2$

Then,

$cd^2 + mnc = n(m^2 + d^2) + m(n^2 + d^2)$

$= mn(m + n) + (m + n)d^2$

$= mnc + cd^2$

This is an identity and no new information is derived.

When d is a median, $m = n = \frac{1}{2}c$. Then

$$cd^2 + \left(\frac{c}{2}\right)\left(\frac{c}{2}\right)c = \left(\frac{c}{2}\right)a^2 + \left(\frac{c}{2}\right)b^2$$

Multiplying out yields:

$4d^2 + c^2 = 2a^2 + 2b^2$

QED

For the case when d is an angle bisector, begin by dividing

$cd^2 + mnc = na^2 + mb^2$ by c and replacing c by m+n in the denominator. This yields

$$d^2 + mn = \frac{na^2 + mb^2}{m+n} \quad \text{and}$$

$$d^2 + mn = \frac{ab\left(\frac{na}{b} + \frac{mb}{a}\right)}{m+n}$$

Specifically, when d is an angle bisector,

$$\frac{m}{a} = \frac{n}{b} \quad \text{or} \quad \frac{m}{n} = \frac{a}{b} \quad \text{or} \quad \frac{n}{m} = \frac{b}{a} \quad *$$

Then we can set $\dfrac{\left(\dfrac{na}{b} + \dfrac{mb}{a}\right)}{m+n} = \dfrac{\left(\dfrac{nm}{n} + \dfrac{mn}{n}\right)}{m+n} = 1$ and

$d^2 + mn = ab$

QED

---

* This assertion was proved in Problem 4C.

Unfortunately, it is rare to see Stewart's Formula discussed, much less proven, in a high school geometry text. Stewart's Formula is the general solution to, perhaps, the most useful, generic triangle situation.

Here, again, we see the power of the Pythagorean Theorem in proving new and varied relationships. The study of cerians is often overlooked in most curricula, but holds a lot of potential for students' independent studies and/or directed projects.

From here, we can explore variations for specific cerians.

The result for altitudes is not surprising since we used the Pythagorean Theorem in deriving Stewart's Formula. I included this case as a warm up for the two that follow. Besides, it is a useful lesson to see that we cannot always expect something new to evolve from every idea we have.

The result for medians has a certain appeal because it resembles the Pythagorean Theorem.

Note that if the original triangle is a right triangle, with $\angle BCA = 90$, then $d = c/2$ (which is another interesting theorem to prove) and our result becomes:

$$4(c/2)^2 + c^2 = 2a^2 + 2b^2 \quad \text{or}$$

$$c^2 = a^2 + b^2$$

Here, again, this result is not surprising since we used the Pythagorean Theorem to prove Stewart's Formula.

If the original triangle is an isosceles right triangle, then d is not only the angle bisector, but also the median and altitude. In that case,

$$d = m = n = c/2 \quad \text{because d is a median.} \qquad \text{Then,}$$

$$d(c/2) + d(c/2) = ab \quad \text{or}$$

$$dc = ab \quad \text{which is true for any right triangle where d is the altitude to the hypotenuse.}$$

The result for angle bisectors agrees with the one we derived using the Law of Cosines in Problem 4C.

7A   Triangle with Equal Altitudes

Prove that if a triangle has two altitudes of equal length, it is an isosceles triangle.

Solution

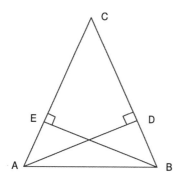

In △ABC, altitudes BE and AD are equal.

In △ACD and △BCE, ∠ADC = ∠CEB because they are both right angles

and ∠ACD = ∠ECB (identity).

Therefore △ACD ~ △BCE because two pairs of angles are equal.

Then ∠CAD = ∠CEB
(If two pairs of angles in a triangle are equal, so is the third pair.)

Adding the given that AD = BE, we get that

△ACD ≅ △ECB (SAS: Side/Angle/Side)

Thus, AC = EC corresponding sides of congruent triangles are equal.

QED

This problem is not an "advanced problem" but it is a "classical problem" in that students have long used the standard SAS proof of congruent triangles for this relatively simple case. This proof is often included in a standard geometry curriculum and is a stepping stone to the more challenging problems to come.

Isosceles triangles have been used for thousands of years in the construction of buildings. The symmetry provides for stability as well as protection from the elements. There are many other applications for isosceles triangles such as design of prisms.

7B   Triangle with Equal Medians

Prove that if a triangle has two medians of equal length, it is an isosceles triangle.

Solution 1

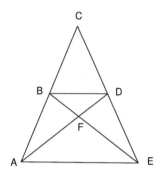

Given: AB = BC, CD = DE and AD = BE.

We see that AF = EF and BF = DF since

medians intersect each other in a 2 to 1 ratio*

and ∠AFB = ∠EFD (vertical angles).

So,   ΔAFB ≅ ΔEFD   (SAS)

Then AB = DE   corresponding parts of congruent triangles are equal.

And AC = CE   doubles of equals are equal.

QED

*   To prove this, draw line BD.

Δ BCD ~ ΔACE and BD = ½ AE and is parallel to AE.

Also, ΔBDF ~ ΔAFE so AF/FD = EF/FB = AE/BD = 2

Solution 2

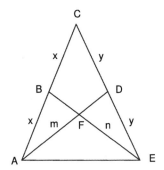

Let AB = BC = x,

CD = DE = y and

AD = m and BE = n

Using the Law of Cosines for $\triangle$ACD,

$$m^2 = (2x)^2 + y^2 - 2(2x)y\cos C = 4x^2 + y^2 - 4xy\cos C$$

Using the Law of Cosines for $\triangle$BCE,

$$n^2 = (2y)^2 + x^2 - 2(2y)x\cos C = 4y^2 + x^2 - 4xy\cos C$$

Since m = n, we can set

$$4x^2 + y^2 - 4xy\cos C = 4y^2 + x^2 - 4xy\cos C$$

which yields $3y^2 = 3x^2$ and x = y.

Then AC = CE.   QED

Two solutions are shown, each more difficult than the one in Problem 7A.

The first solution requires the knowledge that medians intersect each other in a two to one ratio. This is a well known theorem that is often included in geometry texts (and is proven in the solution's footnote). The second solution is a trigonometric one using the Law of Cosines and some algebraic manipulation.

7C   Triangle with Equal Angle Bisectors

Prove that if a triangle has two angle bisectors of equal length, it is an isosceles triangle.

Solution

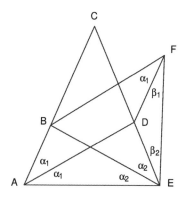

Given $AD = BE$, $\angle CAD = \angle DAE = \alpha_1$ and $\angle CEB = \angle BEA = \alpha_2$.

To prove this theorem, we will need to make additional constructions to the diagram.

> Construct a line parallel to AD through B.
> Construct a line parallel to AB through D.
> Call the intersection of these two lines F and connect FE.

Because ABFD is a parallelogram, $AD = BF$ and $\angle BAD = \angle BFD = \alpha_1$

Then $\angle BFE = \angle BEF$ since base angles of an isosceles triangle are equal.

Or, as shown in the diagram,

$\alpha_1 + \beta_1 = \alpha_2 + \beta_2$

First let us assume that $\alpha_1 > \alpha_2$. The first conclusion, then, is that $\beta_1 < \beta_2$.

If this were so, then $DE < DF$.

It may be intuitive to understand that in a triangle, the leg opposite a larger angle will be larger than a leg opposite a smaller angle. Actually, this is easily proven using the Law of Sines.

Since $DF = AB$ (opposite sides of a parallelogram), $DE < AB$.

Looking at $\triangle ABE$ and $\triangle DEA$, we see that $AE = AE$ (identity) and $BE = AD$ (given). Therefore, if $DE < AB$, then $\angle DAE < \angle BEA$.

It may be intuitive to understand that if two triangles have two equal legs, then the triangle with the larger third leg will also have the larger included angle. Actually, this is easily proven using the Law of Cosines.

If $\angle DAE < \angle BEA$, we can rewrite it as $a_1 < a_2$.

But this contradicts our original assumption.
Therefore, the assumption must be wrong.

Similarly, if we assume $a_1 < a_2$, and follow the same logic, it too will turn out to be contradictory and false.

Therefore, the only remaining possibility is that $a_1 = a_2$.

Then $\angle CAE = \angle CEA$ equals added to equals are equal, and

$CE = AC$ in a triangle, sides opposite equal angles are equal.

QED.

This problem is named the Steiner-Lehmus Theorem and the proof requires a rarely used method called "reductio ad absurdum," Latin for "reduction to absurdity." In practice, it means that if we make an assumption whose resultant consequences are false, then we know that our assumption is also false. Most geometric proofs do not require this approach but, clearly, it was useful here.

Most geometry students learn to include constructions in doing proofs. It is typical to draw altitudes or, perhaps, circles. The proof here requires a construction that is very unusual and creative. One approach would be to offer the students the required construction and then see if they can solve it.

It is also interesting to note that proving a triangle is isosceles given equal altitudes and medians is significantly easier than if given equal angle bisectors.

My all-time favorite teacher, Mr. Jacob Slavin, presented this theorem to the class early in the semester and challenged us to solve it. After the class attempted solutions all semester, he finally shared this proof with us.

8A Inscribed Quadrilateral – Specific Case

Find the radius of a semicircle whose inscribed quadrilateral has side lengths of 1, 2 and 3 and the fourth side is a diameter.

Solution

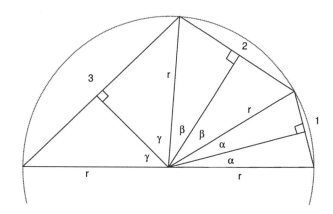

After constructing triangles and altitudes, we see: $\alpha + \beta + \gamma = \pi/2$

Using the geometry, and substituting:

$$\sin^{-1}(\frac{1}{2r}) + \sin^{-1}(\frac{2}{2r}) + \sin^{-1}(\frac{3}{2r}) = \frac{\pi}{2}$$

Transposing the last term on the left and taking the cosine of the equation:

$$\cos[\sin^{-1}(\frac{1}{2r}) + \sin^{-1}(\frac{2}{2r})] = \cos[\frac{\pi}{2} - \sin^{-1}(\frac{3}{2r})]$$

Since $\cos(A+B) = \cos A \cos B - \sin A \sin B$ and $\cos(\frac{\pi}{2} - C) = \sin C$, we get

$$\frac{\sqrt{4r^2 - 1}}{2r} \cdot \frac{\sqrt{4r^2 - 4}}{2r} - \frac{1}{2r} \cdot \frac{2}{2r} = \frac{3}{2r}$$ Multiplying by $4r^2$ and transposing gives

$$\sqrt{4r^2 - 1} \cdot \sqrt{4r^2 - 4} = 6r + 2$$ Then

$4(4r^2-1)(r^2-1) = 4(3r + 1)^2$ and

$4r^4 - 5r^2 + 1 = 9r^2 + 6r + 1$ which reduces to:

$2r^3 - 7r - 3 = 0$

The real root of a cubic equation of the form $x^3 + mx + n = 0$ is:

$$x = 2\sqrt{\frac{-m}{3}} \cos\left(\frac{1}{3}\cos^{-1}\left[\frac{-n}{2}\sqrt{\frac{-27}{m^3}}\right]\right)$$

Substituting $x = r$, $m = -7/2 = -3.5$ and $n = -3/2 = -1.5$ gives

$$r = 2\sqrt{\frac{3.5}{3}} \cos\left(\frac{1}{3}\cos^{-1}\left[\frac{1.5}{2}\sqrt{\frac{27}{42.875}}\right]\right) \qquad \text{or}$$

$r = 2.056546$

When I first tried this problem in geometry class, I got nowhere because I had not yet learned much trigonometry. Later, my uncle, an engineer, suggested the trigonometric approach which led me to the arcsine equation but I did not persist through the necessary expansions. I would return to the problem periodically, and I eventually solved it many years later using the method as shown. Subsequently, I showed this same problem to two college math professors who also independently (and easily) solved the problem as is shown.

This problem clearly requires skills in trigonometry and algebra beyond the standard coursework. It is a good exercise for an in-depth study.

8B Inscribed Quadrilateral – General Case

Find the radius of a semicircle whose inscribed quadrilateral has side lengths of a, b and c and the fourth side is a diameter.

Solution 1

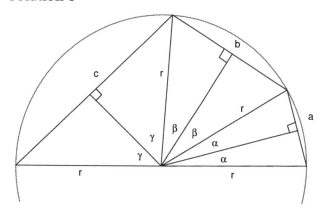

We will use the approach in Solution 1 of Problem 8A.

$$\alpha + \beta + \gamma = \pi/2$$

Using the geometry, and substituting:

$$\sin^{-1}\frac{a}{2r} + \sin^{-1}\frac{b}{2r} + \sin^{-1}\frac{c}{2r} = \frac{\pi}{2}$$

Transposing the last term on the left and taking the cosine of the equation:

$$\cos\left(\sin^{-1}\frac{a}{2r} + \sin^{-1}\frac{b}{2r}\right) = \cos\left(\frac{\pi}{2} - \sin^{-1}\frac{c}{2r}\right)$$

Since $\cos(A + B) = \cos A \cos B - \sin A \sin B$ and $\cos(\frac{\pi}{2}$ - C$) = \sin C$, we get

$$\frac{\sqrt{4r^2 - a}}{2r} \cdot \frac{\sqrt{4r^2 - b}}{2r} - \frac{a}{2r}\cdot\frac{b}{2r} = \frac{c}{2r}$$

Multiplying by $4r^2$, transposing, squaring and simplifying yields,

$$4r^3 - (a^2 + b^2 + c^2)r - abc = 0$$

The real root of a cubic equation of the form $x^3 + mx + n = 0$ is:

$$x = 2\sqrt{\frac{-m}{3}}\cos\left(\frac{1}{3}\cos^{-1}\left[\frac{-n}{2}\sqrt{\frac{-27}{m^3}}\right]\right)$$

Substituting $x = r$, $m = \dfrac{-(a^2+b^2+c^2)}{4}$ and $n = \dfrac{-abc}{4}$ gives

$$r = \sqrt{\frac{a^2+b^2+c^2}{3}}\cos\left(\frac{1}{3}\cos^{-1}\left[abc\sqrt{\frac{27}{(a^2+b^2+c^2)^3}}\right]\right)$$

As a check, substitute: $a = 1$, $b = 2$ and $c = 3$ and get $r = 2.056546$

Solution 2

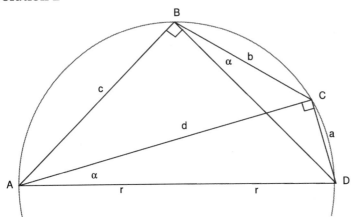

As shown in the diagram:

From $\triangle ACD$, $\sin\alpha = a/(2r)$ and $d^2 = (2r)^2 - a^2 = 4r^2 - a^2$

From $\triangle ABC$, using the Law of Cosines,

$d^2 = b^2 + c^2 - 2bc\cos(90 + \alpha)$

Since $\cos(90 + \alpha) = -\sin\alpha = -\dfrac{a}{2r}$, we can substitute to get

$$d^2 = b^2 + c^2 + \frac{2bca}{2r}$$

We can now set the two expressions for d^2 equal, yielding

$$4r^2 - a^2 = b^2 + c^2 + \frac{2bca}{2r}$$

Multiplying by r and collecting terms, results in

$$4r^3 - (a^2 + b^2 + c^2)r - abc = 0$$

The real root of a cubic equation of the form $x^3 + mx + n = 0$ is:

$$x = 2\sqrt{\frac{-m}{3}}\cos\left(\frac{1}{3}\cos^{-1}\left[\frac{-n}{2}\sqrt{\frac{-27}{m^3}}\right]\right)$$

Substituting $x = r$, $m = \dfrac{-(a^2 + b^2 + c^2)}{4}$ and $n = \dfrac{-abc}{4}$ gives

$$r = \sqrt{\frac{a^2 + b^2 + c^2}{3}}\cos\left(\frac{1}{3}\cos^{-1}\left[abc\sqrt{\frac{27}{(a^2 + b^2 + c^2)^3}}\right]\right)$$

I showed Problem 8A to my friend Dr. Dan Sterling, a former college math professor, who suggested that a more interesting problem would be to parameterize the lengths of the quadrilateral's sides. He then proceeded to solve the problem. Therefore, I would like to propose that the results here be considered a new theorem for quadrilaterals inscribed in a semicircle:

"Sterling's Theorem" – The radius, R, of the circle circumscribing a quadrilateral whose sides are a, b, c and a diameter is:

$$R = \frac{d}{3} \cos\left(\frac{1}{3} \cos^{-1} \frac{abc\sqrt{27}}{d^3} \right) \quad \text{where} \quad d = \sqrt{a^2 + b^2 + c^2}$$

Math teachers debate whether it is better to teach from a specific example to the general case, or vice-versa. Although this text uses the former approach, the latter may be just as useful, or more so.

It would be an interesting experiment to give one set of students Problem 8A, as I first received it, and another set of students "Sterling's Theorem," as stated above. How would we assess the results: quicker solutions, varied approaches, discovering the general case, ability to relate to other principles, etc.?

Problems 8A and 8B are good exercises for an in-depth study because they require some ingenuity to set up the equations and some fortitude to work through the algebra (especially in Solution 1). I should add that both the arcsin equation and the cubic equation, although challenging to solve in closed form, are very easily solved (to about eight significant digits) on a calculator. This is an example, I feel, where using a calculator results in a loss of learning.

While preparing this book, I looked at this problem anew and discovered Solution 2. It isn't the "geometry" solution I was originally seeking, but it is so much simpler and easier. It would be interesting to see how many other solutions have been or can be developed.

Another important point is that both solutions result in the same cubic equation, which cannot be solved by factoring. Since these types of equations are rarely taught in high school courses, the student needs to research the general cubic solution. Since students have studied the Quadratic Formula as the general solution to a second order equation, and now the general solution to a third order equation, they might also be interested in the general solution to a fourth order (quartic) equation.

The general solution of cubic equations can be found on various websites, standard mathematics tables and some math texts. The integration of geometry, trigonometry and algebra in this problem makes this an exciting study option.

9A Factoring a Quartic Binomial – Specific Case

Factor: $x^4 + 4$.

Solution

Notice that $x^4 + 4 = (x^2 + 2)^2 - 4x^2$

This can be written as the difference of two squares:

$x^4 + 4 = (x^2 + 2)^2 - (2x)^2$

which then is factored as

$x^4 + 4 = (x^2 + 2 + 2x)(x^2 + 2 - 2x)$ or, in standard form,

$x^4 + 4 = (x^2 + 2x + 2)(x^2 - 2x + 2)$

This is another problem I learned in high school and have shared with many students throughout the years. It is always a fun brain teaser. The initial approach to use $(x^2 + 2)^2$ is understandable. The key insight is to see that by subtracting off the unwanted $4x^2$, we can create the difference of two squares that factors easily.

Because this is the first official algebra problem, it is appropriate to insert, here, the name of Mohammed ibn Musa al Kwarizmi. His name means: Mohammed, son of Moses, from Kwarizm (a region in modern day Uzbekistan.) He is credited with inventing algebra, named after his term al jabr (meaning "transposing" in Arabic), as well as developing the quadratic formula. He also invented the process of multiplication of Arabic numerals for which the word "algorithm" was created, based on the Latin transliteration of his last name. Living in the ninth century, he calculated, to within one percent accuracy, the circumference of the globe at a time when Europeans still thought the earth was flat.

All high school math students know the names of Euclid and Pythagoras, the most famous contributors to the development of the study of geometry. It is unfortunate that so few students have ever heard of al Kwarizmi, the father of algebra. Math teachers should learn more about the history of this man's work and share it with their students. It makes for a great story and puts a face on a vast body of valuable knowledge.

9B   Factoring a Quartic Binomial – General Case

Find as many methods as possible to factor: $x^4 + y^4$.

Solution 1 A:

Notice that $x^4 + y^4 = (x^2 + y^2)^2 - 2x^2y^2$

This can be written as the difference of two squares:
$x^4 + y^4 = (x^2 + y^2)^2 - (\sqrt{2}\,xy)^2$

which then is factored as
$x^4 + y^4 = (x^2 + y^2 + \sqrt{2}\,xy)(x^2 + y^2 - \sqrt{2}\,xy)$ or, in standard form,

$x^4 + y^4 = (x^2 + \sqrt{2}\,xy + y^2)(x^2 - \sqrt{2}\,xy + y^2)$

Solution 1B:

Using Pythagorean Triplet theory that such triplets can be generated from two integers, m and n, with m > n, by creating three numbers, namely, $m^2 - n^2$, 2mn and $m^2 + n^2$, we can set:

$(m^2 - n^2)^2 = (m^2 + n^2)^2 - (2mn)^2$
$m^4 - 2m^2n^2 + n^4 = (m^2 + n^2)^2 - 4m^2n^2$
$m^4 + n^4 = (m^2 + n^2)^2 - 2m^2n^2$

This can now be factored as the difference of two squares:
$m^4 + n^4 = (m^2 + \sqrt{2}\,mn + n^2)(m^2 - \sqrt{2}\,mn + n^2)$

Solution 2:

Hypothesize that $x^4 + y^4$ can be factored into two standard quadratic forms, with

$x^4 + y^4 = (x^2 + axy + by^2)(x^2 + cxy + dy^2)$

Multiplying out and collecting like terms, yields:

1)	$a + c = 0$	the coefficient of x^3y
2)	$b + d + ac = 0$	the coefficient of x^2y^2
3)	$bc + ad = 0$	the coefficient of xy^3
4)	$bd = 1$	the coefficient of y^4

From 1, a = -c
Substituting a = -c into 3 gives b = d.
Substituting b = d into 4 gives b = d = 1.

(Or, using the symmetry argument: reversing x and y leaves the left side of the equation unchanged, so too would the right side be unchanged, hence b = d = 1.)

Substituting the above into 2 gives a = $\sqrt{2}$ and c = $-\sqrt{2}$.

Then the factors are
$$x^4 + y^4 = (x^2 + \sqrt{2}\,xy + y^2)(x^2 - \sqrt{2}\,xy + y^2)$$

Solution 3

Rewrite $x^4 + y^4$ as the equation $\quad x^4 = -y^4$ and solve using DeMoivre's Theorem.

The roots are:

$x = (y^4)^{1/4}\,[\cos(\pi/4 + n\pi/2) + i\sin(\pi/4 + n\pi/2)]$ for n = 0, 1, 2, 3
in the complex plane.

These can be written as conjugate pairs:
$$x = y \cdot \frac{1 \pm i}{\sqrt{2}} = \frac{(1 \pm i)y}{\sqrt{2}}$$
$$x = y \cdot \frac{-1 \pm i}{\sqrt{2}} = \frac{(-1 \pm i)y}{\sqrt{2}}$$

The first conjugate pair has a sum of $\sqrt{2}$ y and a product of y^2.
The second conjugate pair has a sum of $-\sqrt{2}$ y and a product of y^2.

Therefore, the quadratic factors are:
$$x^4 + y^4 = (x^2 + \sqrt{2}\,xy + y^2)(x^2 - \sqrt{2}\,xy + y^2)$$

Here we have several general case solutions following one special case solution. I have found that the general solutions are more difficult for students to find if they have not seen the specific example first.

Although I have not seen either the specific or the general problem in any high school algebra text, I have seen at least one text that has listed the general solution on a table of polynomial factors. Moreover, current day students are not as familiar with Standardized Math Tables, which almost always have this factor formula in them. Since this is such a great, classic algebra problem, it really ought to be included more frequently.

10A Factoring a Three Variable Quartic Polynomial

Factor: $2a^2b^2 + 2a^2c^2 + 2b^2c^2 - a^4 - b^4 - c^4$

Solution 1

Using the intuitive understanding learned from Problem 9B Solution 1A, we can speculate that there are two polynomial factors that are the difference of two squares.

A reasonable assumption would be to use the form:

$(c^2 - f^2)(g^2 - c^2)$ where f and g are functions of a and b.

This form is selected for the purpose of generating the $-c^4$ term.

Multiplying out yields

$c^2(f^2 + g^2) - f^2g^2 - c^4 = 2a^2b^2 + 2a^2c^2 + 2b^2c^2 - a^4 - b^4 - c^4$ so

$f^2 + g^2 = 2a^2 + 2b^2$ and

$f^2g^2 = -2a^2b^2 + a^4 + b^4 = (a^2 - b^2)^2$

Substituting $g^2 = 2a^2 + 2b^2 - f^2$ into the second equation, yields

$f^2(2a^2 + 2b^2 - f^2) = (a^2 - b^2)^2$ or

$f^4 - (2a^2 + 2b^2)f^2 + (a^2 - b^2)^2 = 0$

Using the quadratic formula to solve for f^2, results in

$f^2 = (a \pm b)^2$

Substituting back yields,

f = a + b   with g = a - b   and

f = a - b with   g = a + b

For the first case,

$(c^2 - f^2)(g^2 - c^2) = [c^2 - (a + b)^2][(a - b)^2 - c^2]$

and, in the second case,

$$(c^2 - f^2)(g^2 - c^2) = [c^2 - (a-b)^2][(a+b)^2 - c^2]$$

We can now factor the differences of two squares easily:

$$[c^2 - (a+b)^2][(a-b)^2 - c^2] = (c+a+b)(c-a-b)(a-b+c)(a-b-c) \qquad \text{and}$$

$$[c^2 - (a-b)^2][(a+b)^2 - c^2] = (c+a-b)(c-a+b)(a+b+c)(a+b-c)$$

We see that both of these answers are identical, so the final answer is:

$$2a^2b^2 + 2a^2c^2 + 2b^2c^2 - a^4 - b^4 - c^4 = (a+b+c)(-a+b+c)(a-b+c)(a+b-c)$$

Solution 2

Perhaps, when we consider factoring $2a^2b^2 + 2a^2c^2 + 2b^2c^2 - a^4 - b^4 - c^4$ our intuitive is not as strong as when we developed Solution 1. We can attempt to use the method in Problem 9B Solution 2.

Let us propose that we can factor the expression into two polynomials in the form:

$$2a^2b^2 + 2a^2c^2 + 2b^2c^2 - a^4 - b^4 - c^4 = (k_1a^2 + k_2b^2 + k_3c^2 + k_4ab + k_5bc + k_6ca)(h_1a^2 + h_2b^2 + h_3c^2 + h_4ab + h_5bc + h_6ca)$$

Multiplying out and collecting and comparing like terms results in 15 equations:

1) $k_1h_1 = -1$
2) $k_2h_2 = -1$
3) $k_3h_3 = -1$
4) $k_4h_1 + k_1h_4 = 0$
5) $k_4h_2 + k_2h_4 = 0$
6) $k_5h_3 + k_3h_5 = 0$
7) $k_5h_2 + k_2h_5 = 0$
8) $k_6h_1 + k_1h_6 = 0$
9) $k_6h_3 + k_3h_6 = 0$
10) $k_2h_1 + k_1h_2 + k_4h_4 = 2$
11) $k_3h_1 + k_1h_3 + k_6h_6 = 2$
12) $k_3h_2 + k_2h_3 + k_5h_5 = 2$
13) $k_4h_3 + k_3h_4 + k_6h_5 + k_5h_6 = 0$
14) $k_5h_1 + k_1h_5 + k_6h_4 + k_4h_6 = 0$
15) $k_6h_2 + k_2h_6 + k_5h_4 + k_4h_5 = 0$

To begin, let us assume $k_1 = 1$. (We will examine this assumption at the end.)

From 1,	$h_1 = -1$
From 8,	$k_6 = h_6$
From 4,	$k_4 = h_4$
From 5,	$k_2 = -h_2$
From 9,	$k_3 = -h_3$
From 6,	$k_5 = h_5$
From 2,	$k_2 = \pm 1$ and $h_2 = \mp 1$
From 3,	$k_3 = \pm 1$ and $h_3 = \mp 1$
From 10,	$(\pm 1)(-1) + (1)(\mp 1) + k_4^2 = 2$
From 11,	$(\pm 1)(-1) + (1)(\mp 1) + k_6^2 = 2$
From 12,	$(\pm 1)(\mp 1) + (\pm 1)(\mp 1) + k_5^2 = 2$ which yields $k_5 = \pm 2$ and $h_5 = \pm 2$
From 14,	$(\pm 2)(-1) + (1)(\pm 2) + k_6 h_4 + k_4 h_6 = 0$ which yields $k_4 = 0$, $k_6 = 0$, or both
From 15,	$k_6(\mp 1) + (\pm 1)k_6 + (\pm 2)k_4 + k_4^2 = 0$ which yields $k_4 = 0$ or $k_4 = \pm 2$
From 13,	$k_4(\mp 1) + (\pm 1)k_4 + k_6(\pm 2) + (\pm 2)k_6 = 0$ which yields $k_6 = 0$

If $k_4 = 0$, from 10, we get $k_2 = -1$ and $h_2 = 1$

If $k_6 = 0$, from 11, we get $k_3 = -1$ and $h_3 = 1$.

The results then become:

$$2a^2b^2 + 2a^2c^2 + 2b^2c^2 - a^4 - b^4 - c^4 = (a^2 - b^2 - c^2 + 2bc)(- a^2 + b^2 + c^2 + 2bc)$$

We can now write these factors as the differences of squares:

$$(a^2 - b^2 - c^2 + 2bc)(- a^2 + b^2 + c^2 + 2bc) = [a^2 - (b - c)^2][(b + c)^2 - a^2]$$

$$= (a + b - c)(a - b + c)(b + c + a)(b + c - a) \qquad \text{and}$$

$$2a^2b^2 + 2a^2c^2 + 2b^2c^2 - a^4 - b^4 - c^4 = (a + b + c)(- a + b + c)(a - b + c)(a + b - c)$$

If we reexamine our assumption that $k_1 = 1$, we see that if we had chosen $k_1 = -1$, the process would have unfolded in the same manner and the result would have been the same.

This factoring problem can be considered a lemma for the solution of Problem 5A. Both Solutions 1 and 2 rely on some level of intuition or creativity in setting up the format of the approach. It is precisely this kind of perspicacity that serves the serious math student.

An important note about the solution is that because there are four unique trinomial factors, there are three different ways they can be paired together. That means that the two quadrinomial quadratic factors are not unique. This is seen in Solution 1 when we start with separating c from a and b, thereby generating two of the three quadratic factors. In Solution 2, the selection of k_1 results in the separation of a from b and c. For $k_1 = -1$, the quadratic factors, again, will be different.

11A   Square Rooting - Complex Numbers

Find the square roots of complex number a + bi.

Solution 1

Assume $\sqrt{a+bi} = c + di$

Then $a + bi = c^2 - d^2 + 2cdi$

We can set the coefficients of both the real and imaginary terms equal to each other. Then

$c^2 - d^2 = a$ and

$2cd = b.$ Then

$c = \dfrac{b}{2d}$ and

$\dfrac{b^2}{4d^2} - d^2 = a$ This becomes

$4d^4 + 4ad^2 - b^2 = 0$ We can solve for d^2 using the quadratic formula.

$d^2 = \dfrac{-4a \pm \sqrt{16a^2 + 16b^2}}{8} = \dfrac{-a \pm \sqrt{a^2 + b^2}}{2}$

$d = \dfrac{1}{\sqrt{2}}\sqrt{\sqrt{a^2 + b^2} - a}$

After solving* for d, substitute in $c = \dfrac{b}{2d}$ for c.

$c = \dfrac{b}{2d} = \dfrac{b}{\sqrt{2(\sqrt{a^2+b^2} - a)}}$

$2c^2(\sqrt{a^2+b^2} - a) = b^2$

$\sqrt{a^2+b^2} = a + \dfrac{b^2}{2c^2}$

$a^2 + b^2 = a^2 + \dfrac{ab^2}{c^2} + \dfrac{b^4}{4c^4}$

$4c^4 - 4ac^2 - b^2 = 0$

$$c^2 = \frac{4a \pm \sqrt{16a^2 + 16b^2}}{8} = \frac{a \pm \sqrt{a^2 + b^2}}{2} \quad \text{and}$$

$$c = \frac{1}{\sqrt{2}}\sqrt{\sqrt{a^2 + b^2} + a}$$

(Or, using the symmetry argument for the equations $c^2 - d^2 = a$ and $2cd = b$, interchanging c and d replaces a by $-a$ but leaves b unchanged. Therefore, since

$$d = \frac{1}{\sqrt{2}}\sqrt{\sqrt{a^2 + b^2} - a}, \quad c = \frac{1}{\sqrt{2}}\sqrt{\sqrt{a^2 + b^2} + a} \,)$$

Example: Find $\sqrt{3 + 4i}$

Here, a = 3 and b = 4.
$$d^2 = \frac{-3 \pm \sqrt{9 + 16}}{2} = -4 \text{ and } 1.$$

Since d is real, d = 1 or -1 and c = 4/2 = 2 or c = 4/(-2) = -2

Check: (2 + i)(2 + i) = 4 -1 + 4i = 3+ 4i
        also (- 2 – i)(- 2 – i) = 3 + 4i

So $\sqrt{3 + 4i} = \pm(2 + i)$

And, even though d is defined as real, if we take the alternate choice for d^2, that is $d^2 = -4$,

we get d = ±2i. Then $c = \dfrac{b}{2d} = \dfrac{4}{\pm 4i} = \mp i$

Thus: $\sqrt{(3 + 4i)} = \mp i \pm 2i \cdot i = \pm(2 + i)$ which is the same answer as for $d^2 = 1$.

* This process may require taking the square root of a mixed rational and irrational number which is addressed in Problem 11B.

Solution 2

Using the polar coordinate representation of a complex number, we can set

$$a+bi = \sqrt{a^2+b^2} \cdot \left[\cos(\tan^{-1}\frac{b}{a}) + i \cdot \sin(\tan^{-1}\frac{b}{a}) \right]$$

Then, $\sqrt{a+bi} = \sqrt[4]{a^2+b^2} \cdot \left[\cos(\frac{1}{2}\tan^{-1}\frac{b}{a}) + i \cdot \sin(\frac{1}{2}\tan^{-1}\frac{b}{a}) \right]$

Since $\cos(\tan^{-1}\frac{b}{a}) = \dfrac{a}{\sqrt{a^2+b^2}}$

and $\cos\dfrac{\theta}{2} = \sqrt{\dfrac{1+\cos\theta}{2}}$ (the half angle formula)*

$$\cos\left[\frac{1}{2}\tan^{-1}\frac{b}{a}\right] = \frac{(\sqrt{a^2+b^2}+a)^{\frac{1}{2}}}{\sqrt{2} \cdot \sqrt[4]{a^2+b^2}}$$

Similarly,

$$\sin\frac{\theta}{2} = \sqrt{\frac{1-\cos\theta}{2}}$$ (again, the half angle formula)*

then $\sin(\frac{1}{2}\tan^{-1}\frac{b}{a}) = \dfrac{(\sqrt{a^2+b^2}-a)^{\frac{1}{2}}}{\sqrt{2} \cdot \sqrt[4]{a^2+b^2}}$

Putting the elements together,

$$c = \frac{1}{\sqrt{2}}\sqrt{\sqrt{a^2+b^2}+a} \quad \text{and} \quad d = \frac{1}{\sqrt{2}}\sqrt{\sqrt{a^2+b^2}-a}$$

* The half angle formula for cosine and sine include a $\pm\sqrt{}$ function depending on the angle's quadrant. This derivation only addresses angles in the first quadrant.

This problem opens up a very important topic. When students are first introduced to imaginary numbers, teachers are often challenged to explain them in "realistic" terms. How can they be made "real?"

One answer is to share both the history and geometric interpretation of imaginary numbers. Historically, imaginary numbers arose as the solutions to quadratic and cubic equations, but even mathematicians had a hard time believing they really existed.

Probably, the best explanation is to suggest that the real number line (x – axis) gets rotated 180 degrees counterclockwise when multiplied by (-1). Since i times i is (-1), multiplying by i is like rotating only 90 degrees. Now the real number line has been converted into the complex number plane and all the algebraic properties of complex numbers (such as addition, subtraction, multiplication, etc) can be shown to have geometric interpretations and analogies. This helps make imaginary numbers real.

Introducing students to the engineering use of imaginary numbers is another good approach. For example, electric circuit analysis techniques use imaginary numbers to model capacitors and inductors in the same way that resistors are modeled. This unifies and empowers the entire field of electrical engineering. (Interestingly, in electric circuit analysis the letter j is used, rather than i; that is, $j = \sqrt{-1}$. This is due to the convention of using i to represent the electric current in the circuit.) See Problem 14A.

11B Square Rooting - Irrational Numbers

Find the square root of a mixed irrational number $a \pm \sqrt{b}$.

Solution

The solution here is similar to Solution 1 of Problem 11A.

Let $\sqrt{a \pm \sqrt{b}} = c \pm d\sqrt{b}$ Then,

$a \pm \sqrt{b} = c^2 + d^2 b \pm 2cd \sqrt{b}$

Equating like terms gives $a = c^2 + d^2 b$ and $1 = \pm 2cd$

With $c = \pm \dfrac{1}{2d}$ and $c^2 = \dfrac{1}{4d^2}$

$a = \dfrac{1}{4d^2} + d^2 b$

$4bd^4 - 4ad^2 + 1 = 0$ Solve for d^2 using the Quadratic Formula

$d^2 = \dfrac{a \pm \sqrt{a^2 - b}}{2b}$ Then

$d = \dfrac{1}{\sqrt{2b}} \sqrt{a \pm \sqrt{a^2 - b}}$ and

$c = \pm \dfrac{1}{2d} = \pm \dfrac{1}{\sqrt{2}} \sqrt{a \mp \sqrt{a^2 - b}}$

Example: For $a = 3$ and $b = 5$

$d^2 = \dfrac{3 \pm \sqrt{9-5}}{10} = \dfrac{1}{2}$ and $\dfrac{1}{10}$

For $d^2 = \frac{1}{2}$, $d = \pm \dfrac{1}{\sqrt{2}}$ and $c = \pm \dfrac{\sqrt{2}}{2} = \pm \dfrac{1}{\sqrt{2}}$

For $d^2 = \dfrac{1}{10}$, $d = \pm \dfrac{1}{\sqrt{10}}$ and $c = \pm \dfrac{\sqrt{10}}{2}$

Both sets of values yield the same results, namely $\sqrt{3 + \sqrt{5}} = \pm(\dfrac{1}{\sqrt{2}} + \dfrac{\sqrt{5}}{\sqrt{2}})$

The advent of electronic calculators has blunted the efforts of teachers to study these kinds of problems. Even more so, calculators have blunted the interest of students to learn these kinds of solutions. Still, these are valuable exercises for the serious student.

Some math teachers believe that using calculators and software programs are the modern, productive and efficient way to learn math. Certainly, high tech tools have their place and are used extensively in industry. I would still contend that deeper understanding is more likely to come by not using these automation tools, but by thinking through the principles involved in a problem and carrying out the logic algebraically. (Then use the calculator if it is needed.)

Slide rules were the best tools we used to have; but note that a slide rule could not solve these kinds of problems. Sometimes, I bring my old slide rules to class to show my students what life was like "back then." They are surprised to see that certain calculations can be done on a slide rule as fast as they can be done on a calculator.

12A The Irrationality of the Square Root of Two

Prove that $\sqrt{2}$ is irrational.

That is, show that $\sqrt{2}$ is not equal to the quotient of two integers.

Solution

Let us assume that $\sqrt{2}$ is a rational number equal to a/b where the fraction has been reduced and there are no other common factors between a and b other than 1.

Then $\sqrt{2} = \dfrac{a}{b}$.

Squaring both sides yields
$2 = \dfrac{a^2}{b^2}$ and

$a^2 = 2b^2$

This means that a must be an even number and we can set a = 2c.

Then $(2c)^2 = 2b^2$

or $4c^2 = 2b^2$

and $2c^2 = b^2$

This would mean that b is an even number as well.

Our assumption that $\sqrt{2}$ is a rational number equal to $\dfrac{a}{b}$, where the fraction has been reduced and there are no other common factors between a and b other than 1, has led to a contradiction that both a and b are even and have the factor 2 in common.

Therefore, our assumption is false and the conclusion is that $\sqrt{2}$ is irrational.

QED

Square roots were very important years ago because land was owned in square lots. "Square roots" were often referred to as "sides" of a number. That is $\sqrt{2}$ (or side of 2) was the length of a plot that would be twice as large as a standard size plot of unit area.

Hopefully, the comments in Problem 2A, indicating the peril of sharing this 2500 year old secret, has not deterred any students. Having reached Problem 12A, the student should feel safe enough to give this problem a try.

The elegance of the logic of this proof leaves us with a sense of awe and appreciation for our forebears so long ago. Recreating this proof puts us in the shoes of those who walked millennia ago.

Note also, that again, we used the reductio ad absurdum method of proof as we did in Problem 7C.

13A   A Quartic Equation – Specific Case

Solve for x: $x^4 + 5x^3 + 6x^2 + 5x + 1 = 0$.

Solution

The key to solving this quartic equation is to see the symmetry of the coefficients, divide the equation by x^2 and regroup the terms. The result is:

$$(x^2 + \frac{1}{x^2}) + 5(x + \frac{1}{x}) + 6 = 0$$

Now let $x + \frac{1}{x} = y$

then $x^2 + \frac{1}{x^2} = y^2 - 2$

Substituting gives us

$y^2 - 2 + 5y + 6 = 0$ or

$y^2 + 5y + 4 = 0$

Ordinarily, we could always solve using the Quadratic Formula, but here we can factor.
$(y + 4)(y + 1) = 0$

$y = -4$ and $y = -1$.

Now, solving for x,

$$x + \frac{1}{x} = -4 \quad \text{and} \quad x + \frac{1}{x} = -1$$

These two quadratic equations yield the four answers for x, namely:

$x^2 + 4x + 1 = 0$ yields

$x = -2 \pm \sqrt{3}$ and

$x^2 + x + 1 = 0$ yields

$x = \dfrac{-1 \pm i\sqrt{3}}{2}$

Quartic (and cubic) equations are rarely discussed in a math curriculum but the standard form solutions are available in tables for those who seek them. The third and fourth order equations that are discussed are usually easily factorable because they are missing terms, they are perfect squares or they are the difference of squares.

The student can try figuring out the quadratic factors through trial and error, or perhaps another more logical approach. And, there is always the general quartic solution to fall back on. For this problem the solution lies in the equation's symmetry.

As mentioned, solving for x may require finding the square root of a complex and/or irrational number. At that point, another level of skill is required.

Here, in the Algebra section, having solved a quartic equation using factoring, it seems like a good place to talk about factoring, solving single and simultaneous equations, and algebra in general.

Most high school graduates do not consciously use algebra in their daily lives although they often solve "word problems" without quite realizing it. Students, who develop strong algebra skills, find it very handy when working on personal finances as well as home construction projects, etc. Almost all technological products such as TV's, telephones, airplanes, cars, computers, etc were designed and/or operate on the basis of algebraic equations. Business transactions, taxes, insurance and government programs all rely on algebra to implement their objectives.

Computer Aided Design (CAD) using Finite Element Analysis (FEA) is most commonly used in mechanical engineering tasks to simulate and analyze almost any animate or inanimate object, and is frequently used in animated film production. FEA takes complex shapes, simplifies them into many discrete, interconnected parts and uses simultaneous algebraic equations to figure out how they all work together. Similarly, Magnetic Resonance Imaging (MRI) uses algebra to get a clear picture of how our body parts are positioned relative to each other.

Algebra, and mathematics in general, can be considered tools in a toolbox that can be pulled out to find a needed answer. If the tool is not available, the solution may be harder to find. The public at large never uses a stethoscope, a scalpel or an X-ray machine themselves, but they are aware of these tools and how helpful they can be in the hands of a practitioner. It is good to see that, in recent years, the public has also come to appreciate the benefits of math, such as algebra and geometry, when used as tools by a technologist. Improving high school math education helps bring this awareness to everyone, both the students who will be using advanced math in their careers and the students who think they won't ever use it again.

13B   A Quartic Equation – General Case

Solve for x in the general case of the special quartic equation:

$$x^4 + ax^3 + bx^2 + ax + 1 = 0.$$

Solution

Following the procedure outlined in Problem 13A

$$(x^2 + \frac{1}{x^2}) + a(x + \frac{1}{x}) + b = 0$$

Now let $\qquad x + \dfrac{1}{x} = y$

then $\qquad x^2 + \dfrac{1}{x^2} = y^2 - 2$

Substituting gives us

$$y^2 + ay + (b - 2) = 0$$

Using the quadratic formula to solve for y,

$$y = \frac{-a \pm \sqrt{a^2 - 4(b-2)}}{2}$$

Solving for x,

$$x^2 - yx + 1 = 0 \qquad\qquad \text{and}$$

$$x = \frac{y \pm \sqrt{y^2 - 4}}{2}$$

The solution for y can be irrational and/or imaginary. Then the solution for x will require taking the square root of an irrational and/or imaginary number, the process for which has been discussed in Problems 11A and 11B.

In between "easy" quartics and the standard form quartics are those with symmetric coefficients. As we have seen, there is a straight-forward solution available when the symmetry is put to good use. There is a simple elegance to the symmetry and the solution, at least up to the solution for x^2.

In this case, the general solution may be easier to solve than the specific case because the symmetry is more obvious. Still, most students would not be looking to divide by x^2 and then making the appropriate substitution. It may be, though, that using the general case equation, rather than the specific one, will help the students to think mathematically. The question would make for an interesting controlled experiment.

Here, again, it is recognized that there may be several other solutions available via the imagination and creativity of the students.

14A   Applying Imaginary Numbers

For the electric circuit shown, derive the amplitude, a, and phase, ϕ, of the output signal as a function of input voltage frequency, ω. Also find ω_0, the frequency that maximizes the amplitude, and ω_1 and ω_2, the frequencies at which the amplitude is $\frac{\sqrt{2}}{2}$ of its maximum. Lastly, find the bandwidth, $\omega_2 - \omega_1$.

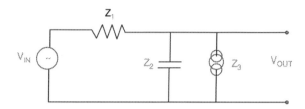

Given:

$$V_{IN} = \sin \omega t$$

$$V_{OUT} = a \sin(\omega t + \phi)$$

$$a = \frac{V_{OUT}}{V_{IN}} = \left([\mathrm{Re}\frac{Z_P}{Z_1 + Z_P}]^2 + [\mathrm{Im}\frac{Z_P}{Z_1 + Z_P}]^2 \right)^{\frac{1}{2}}$$

$$\phi = \tan^{-1}\left(\frac{\mathrm{Im}\dfrac{Z_P}{Z_1 + Z_P}}{\mathrm{Re}\dfrac{Z_P}{Z_1 + Z_P}} \right)$$

where

$$Z_1 = R$$

$$Z_2 = \frac{1}{j\omega C}$$

$$Z_3 = j\omega L$$

$$Z_P = \frac{Z_2 Z_3}{Z_2 + Z_3} \quad \text{is the combined impedance of } Z_2 \text{ (the impedance of the capacitor) and}$$

Z_3 (the impedance of the inductor) connected in parallel, and Z_1 is the impedance of the resistor.

Note: $j = \sqrt{-1} = i$ (which is the usual symbol for the imaginary number unit).

124

Solution

$$Z_P = \frac{Z_2 Z_3}{Z_2 + Z_3} = \frac{\dfrac{1}{j\omega C} \cdot j\omega L}{\dfrac{1}{j\omega C} + j\omega L} = \frac{j\omega L}{1 - \omega^2 LC}$$

$$Z_1 + Z_P = R + \frac{j\omega L}{1 - \omega^2 LC} = \frac{R(1 - \omega^2 LC) + j\omega L}{1 - \omega^2 LC}$$

$$\frac{Z_P}{Z_1 + Z_P} = \frac{\dfrac{j\omega L}{1 - \omega^2 LC}}{\dfrac{R(1 - \omega^2 LC) + j\omega L}{1 - \omega^2 LC}} = \frac{j\omega L}{R(1 - \omega^2 LC) + j\omega L}$$

$$\frac{Z_P}{Z_1 + Z_P} = \frac{j\omega L}{R(1 - \omega^2 LC) + j\omega L} \cdot \frac{R(1 - \omega^2 LC) - j\omega L}{R(1 - \omega^2 LC) - j\omega L}$$

$$\frac{Z_P}{Z_1 + Z_P} = \frac{\omega^2 L^2 + j\omega LR(1 - \omega^2 LC)}{\omega^2 L^2 + R^2(1 - \omega^2 LC)^2}$$

$$a = \frac{\sqrt{\omega^4 L^4 + \omega^2 L^2 R^2 (1 - \omega^2 LC)^2}}{\omega^2 L^2 + R^2 (1 - \omega^2 LC)^2}$$

$$\phi = \tan^{-1}\left(\frac{\omega LR(1 - \omega^2 LC)}{\omega^2 L^2}\right) = \tan^{-1}\left(\frac{R(1 - \omega^2 LC)}{\omega L}\right)$$

To maximize a, the condition that $\omega = \omega_0$,
$1 - \omega_0^2 LC = 0$

$$\omega_0 = \frac{1}{\sqrt{LC}}$$

and a = 1.

For $a = \dfrac{\sqrt{2}}{2}$, $\operatorname{Re}\left(\dfrac{Z_P}{Z_1 + Z_P}\right) = \operatorname{Im}\left(\dfrac{Z_P}{Z_1 + Z_P}\right)$ and $\phi = \pm\dfrac{\pi}{4}$

If we select $\omega = \omega_2 > \omega_0$, then $\phi = -\dfrac{\pi}{4}$ and $\tan\phi = -1$.

Setting $\omega_2 = k_2 \omega_0$, with $k_2 > 1$,

$$-1 = \frac{R(1 - k^2 \omega_0^2 LC)}{k\omega_0 L} \quad \text{and} \quad k_2^2 \omega_0^2 RLC - k_2 \omega_0 L - R = 0.$$

Substituting: $LC = \dfrac{1}{\omega_0^2}$ and $L = \dfrac{1}{\omega_0}\sqrt{\dfrac{L}{C}}$ we get

$$k_2^2 R - k_2 \sqrt{\frac{L}{C}} - R = 0 \quad \text{and} \quad k_2^2 - \frac{1}{R}\sqrt{\frac{L}{C}} k_2 - 1 = 0.$$

If we set $Q = R\sqrt{\dfrac{C}{L}}$, then $k_2 = \dfrac{1}{Q}\left[\dfrac{1+\sqrt{4+Q^2}}{2}\right]$.

Similarly, if we select $\omega = \omega_1 < \omega_0$, then $\phi = +\dfrac{\pi}{4}$ and $\tan\phi = +1$.

Setting $\omega_1 = k_1\omega_0$, with $k_1 < 1$, we get $k_1 = \dfrac{1}{Q}\left[\dfrac{-1+\sqrt{4+Q^2}}{2}\right]$.

Finally, the bandwidth is $\omega_2 - \omega_1 = \dfrac{1}{Q}\omega_0$. Note that for the special case where Q=1

$$k_2 = \frac{1}{k_1} = \frac{1+\sqrt{5}}{2} \approx 1.618 \quad \text{and} \quad \omega_2 - \omega_1 = \omega_0.$$

One of the most glaring shortcomings of a strong high school math program is the lack of even one practical, real life application of the use of imaginary numbers. Typically, students are given many exercises to calculate the sum, difference, product, quotient and exponents of complex numbers and the grand finale is learning to use DeMoivre's Theorem and/or Euler's Formula. No example is offered that refers to an application situation. I have discussed this with math teachers and reviewed many texts and their conclusion is to just teach without any.

Problem 14A is a circuit design and analysis example from the daily life of electrical engineers that I have used successfully in my classes. (Since electrical engineers reserve the letter i to represent electric current, they use the letter j as the imaginary unit, as we have done here.) That I worked these problems as a third year engineering student in college only further inspires the students to get involved with it. Having the students work a "real" problem responds to those complaints that imaginary numbers are an irrelevant waste of time.

Capacitors, which store electric energy, and inductors, which store magnetic energy, can be represented mathematically as imaginary resistors, which are conductors. This representation is derived from differential equations that model the physics of these components (and as such does not have to be presented to the student). The point is that circuits can be designed easily, using imaginary numbers, rather than having to solve very difficult sets of simultaneous differential equations.

As described, this circuit is an oscillator and resonates with a frequency of $\omega_0 = 1/\sqrt{LC}$. That is, the amplitude or gain of the circuit is its maximum (a = 1) and the phase shift is zero at the resonant frequency. Above or below the resonant frequency, the circuit attenuates the incoming signal and phase shifts it out of synchronization. The circuit is commonly used to tune into individual radio or TV stations, allow both voice and DSL communication on a single phone line, and many other applications. The formulas for calculating amplitude and phase are derived from a basic application of Ohm's Law for series and parallel circuits, which is taught in high school physics.

The frequencies, ω_1 and ω_2, where $a = \sqrt{2}/2$ were selected because it is at this point that only half the power of the input signal makes it through the circuit. Above or below these frequencies, the signals are significantly attenuated which is why the circuit is called a "band-pass filter" and the "bandwidth" is defined at these point. As shown, the bandwidth defining frequencies can be calculated in terms of the circuit elements.

The big surprise is that when $Q = 1$, ω_1 and ω_2 are in relationship to ω_0 as defined by the Golden Ratio! It is amazing how this ratio keeps coming up. I do not believe there are any engineering texts that describe this phenomenon. When one of my classes was working on the circuit problem, I brought in students from another class who were working on the Golden Ratio, so they could act as consultants. Neither set of students knew what was going to happen until the results appeared. It was a great synergy.

The principle of using imaginary numbers is also applied to other areas of electrical engineering such as in control systems design and analysis. Here, differential equations model that which is to be controlled, such as an aircraft, and Laplace Transforms use the complex number plane to model an entire feedback control system using rational functions. Amplitude and frequency plots, such as in the electric circuit problem, are used to design electric circuit compensators that provide rapid system response to environmental disturbances as well as stability over a wide range of operating conditions.

Since our class topic following imaginary numbers was rational functions, which are also taught without practical examples, I used the opportunity to present an aircraft control system design lecture. The class had a very positive response, not only because they could follow along with such a complex and interesting subject but for three other reasons: they liked the idea that this material came from a senior year course in engineering school (and asked that I now confer a college degree on them); they saw how an engineer really earns a good salary; and since the same processes are used for all aerospace vehicles, they learned what a rocket scientist really does.

The algebraic manipulation of the imaginary numbers is a worthy exercise for the math student, even without any interest in the physics of the situation. Transforming a fraction with a complex numerator and denominator into standard $a + bi$ form by multiplying numerator and denominator by the conjugate of the denominator is a key technique to learn. Knowing early in their technical careers that there truly are practical and important uses for imaginary numbers adds power to the students' study of imaginary numbers. Imaginary numbers are used in the technical analysis of virtually all oscillatory motion found in nature, including: mechanical vibration, structural resonance, hydrodynamics, quantum mechanics, etc. Students can know that imaginary numbers are "real."

15A   Imaginary Exponents

Using Euler's Formula: $e^{i\theta} = \cos\theta + i\sin\theta$,
calculate the value of a^i and i^i, where a is a real number and $i = \sqrt{-1}$.

Solution

To find a^i, we know that

$a = e^{\ln a}$ so

$a^i = e^{i\ln a}$

Using Euler's Formula

$e^{i\theta} = \cos\theta + i\sin\theta$ where $i = \sqrt{-1}$.
we get

$a^i = \cos(\ln a) + i\sin(\ln a)$

To find i^i, we again use Euler's Formula: $e^{i\theta} = \cos\theta + i\sin\theta$
and select the case where $\theta = \pi$.

Since $\cos \pi = -1$ and $\sin \pi = 0$,

$e^{i\pi} = -1$

We can substitute i^2 for -1 and get

$e^{i\pi} = i^2$

Let us now raise this equation to the i power. We calculate the new exponents by multiplying the original exponent by i.   That is,

$(e^{i\pi})^i = (i^2)^i$ and

$e^{-\pi} = i^{2i}$

Taking the square root of the equation yields

$e^{-\pi/2} = i^i$ or

$i^i = e^{-\pi/2} \approx 0.20788$

That a real number raised to an imaginary exponent is a complex number may seem intuitively correct since we start with a real part and an imaginary part. Euler's Formula quantifies the relationship and you can even do the calculation on a calculator.

But, isn't it amazing that i to the i power is a real number and is approximately equal to 1/5?

The scientist, Richard Feynman, once called $e^{i\pi} = -1$ "the most remarkable formula in math," perhaps because it relates real, imaginary and transcendental numbers together in a simple package. The idea that an imaginary number to an imaginary power is real is truly astounding.

Considering that 1 to the 1 power is 1, it is even more astounding that i to the i power is about 0.2. If that is not enough, try calculating i to the i power on a calculator. It gives the correct answer!

This problem is an advanced one and is based on previous knowledge about the number e, Euler's constant. This is a transcendental number, as is π, and the value is approximately 2.718... Transcendental numbers are numbers that are not derived as solutions to polynomial equations with real coefficients. Ordinarily, students first encounter e in the study of derivatives during the first semester of calculus. In searching for a function whose derivative is the same as the function itself, the field is narrowed to exponential functions of the form a^x, but their derivatives always have an extra coefficient when differentiated. When a equals e, the result is pure: the derivative and the function are the same.

Suffice it to say that the number of examples of the functionality of e^x in nature is virtually innumerable. The growth of life, the decay of atoms, the compounding of interest, the transfer of heat, the dampening of motion, etc are only a short list. Math teachers who can use the pedagogy of projects can find many topics that will engage students with results that the student will remember for life.

Of the many topics that can be discussed, here are two that work well.

The first begins by asking, why does a tree look like a tree? The initial response is usually a perplexed one. Then, discuss how at any height above the ground, the tree has to support the weight above that point. This leads to a simple differential equation whose solution is the form of e^x. Look at the profile of a tree: it is in the shape of e^x. A reading of the poem "Trees" by Joyce Kilmer is a nice ending to the lesson.

The second begins by asking, who would like to learn how to make a million dollars? Here, no one is perplexed; everyone is with you. Then, the lesson makes exponents and logarithms very important tools in choosing money making strategies. Semi-log paper is another good tool to introduce here.

16A Probability

Our intrepid statistics student, Dewey Cheatham, was planning to take a test in his literature class and he had heard a rumor that it would be 20 True/False questions. Because he had not read the required reading, and had no clue at all what the book was about, his plan was to take random guesses for each answer. Dewey figured that on average, he should get 50% correct and that his chances of getting a passing grade of at least 60% were pretty good. Unfortunately, when he actually took the exam, he saw that there were, in fact, 20 multiple choice questions, each with five possible answers. Needless to say, Dewey realized that by choosing his answers randomly, his chances for getting a passing grade went down considerably.

Using random guessing, what were Dewey's chances of passing with at least a 60% grade in the case of 20 True/False questions? What were they in the case of 20 five option multiple choice questions?

Solution

For n independent events, each with a probability of success of p, the probability of getting exactly r successes is $p^r (1-p)^{(n-r)}$ for any given sequence. Since we are not concerned about the sequence of successes, the probability goes up by a factor of $_nC_r$ where $_nC_r$ is the number of combinations of n things taken r at a time.

$_nC_r$ is calculated by $_nC_r = \dfrac{n!}{(n-r)!\,r!}$

In the first case, n = 20, p = 0.5 (because guessing True/False questions correctly is a 50/50 chance) and r = 60% of 20 = 12.

Therefore, the probability of getting exactly 12 correct answers is:
$_{20}C_{12}(0.5)^{12}(1-0.5)^8 = .1201$

Since Dewey would be even more pleased with a higher score, we calculate the probabilities for r equal to 13, 14, ... 20. These are:

$_{20}C_{13}(0.5)^{13}(1-0.5)^7 = .0739$

$_{20}C_{14}(0.5)^{14}(1-0.5)^6 = .0370$

$_{20}C_{15}(0.5)^{15}(1-0.5)^5 = .0148$

$_{20}C_{16}(0.5)^{16}(1-0.5)^4 = .0046$

$_{20}C_{17}(0.5)^{17}(1-0.5)^3 = .0011$

$_{20}C_{18}(0.5)^{18}(1-0.5)^2 = .0002$

The results for r = 19 and r = 20 are insignificantly small.

Then the total probability of receiving a grade of at least 60% is approximately:
.1201 + .0739 + .0370 + .0148 + .0046 + .0011 + .0002 = .2517 or 25.17%.

This can also be calculated more directly since p = 1 - p. For each r, the probability is equal to $_nC_r(0.5)^{20}$.

where $(0.5)^{20} = \dfrac{1}{1,048,576}$.

In other words, the probability of r successes is equal to the number of combinations of n things taken r at a time divided by the total number of possible outcomes which, in this case, is 2^{20} or 1,048,576.

In the scenario where Dewey's exam was 20 five-option multiple choice questions, p becomes 0.2 and 1 - p = 0.8.

Then, the probability of exactly 12 correct answers is

$$_{20}C_{12}(.2)^{12}(0.8)^8 = .0000866$$

The probability of exactly 13 correct answers is

$$_{20}C_{13}(.2)^{13}(.8)^7 = .0000133 \text{ and}$$

The probability of exactly 14 correct answers is

$$_{20}C_{14}(0.2)^{14}(0.8)^6 = .0000017$$

The probabilities of 15 or more correct answers are insignificantly small.

Therefore, the probability of Dewey's receiving a grade of at least 60% is approximately

.0000866 + .0000133 + .0000017 = .0001016 or about 0.01%.

The first thing that jumps out about this problem is that it is written in a format and a style that are much different from all the preceding problems. The format is called a "word problem" and the style is more conversational, informal and, sometimes, humorous.

Word problems have often been the most difficult for students, even those who do well with the "regular" problems. A word problem does not give directions as explicitly as "prove the Pythagorean Theorem" or "factor $x^4 + 4$". Word problems require the skill to take the words, create diagrams, mathematical representations, equations, etc. This is a particularly more difficult concept to teach. Perhaps, that is why math teachers allow themselves the liberty of injecting some humor into the problem in an attempt to make the environment more congenial for students.

This problem is a probability problem that is on the boundary with the field of study of statistics. The boundary is a gray line at best, so this problem is the first of the statistics problems.

There is an old saying that one has to watch out for lies, damn lies and statistics. Statistics can be used as a powerful tool of the unscrupulous and/or uninformed. It is easy to make things seem as they are not. In the study of statistics there are two kinds of errors: Type I where you believe, or accept, a conclusion that is not true; and Type II where you do not believe, or accept, a conclusion that is true. Either error can be very costly. Dewey Cheatham is my symbolic character that represents the sometime dubious ethics of users of statistics. On the other hand, he's probably a good statistician. I tell my Statistics students that I want them to learn how to lie, cheat and steal (using statistics) so that they will be able to identify those behaviors in others and, thereby, be prepared.

The problem, itself, is a striking example of how some complex math can clearly describe a simple and practical situation. And, certainly, the use of a calculator comes in mighty handy to perform the numerical operations required.

It should also be said that the assumption that a student can perform random guessing is, in fact, the least plausible part of the story. Generating random True/False answers or random a, b, c, d, e answer choices is not as easy as it seems. Usually, students will follow a pattern which, ultimately, is statistically no better or worse than any other guessing method. The second least plausible part of the story is that Dewey's assumption he will get 50% of the True/False questions correct is only valid if he takes a test with hundreds of questions (if not thousands). With only 20 questions, Dewey can be about 93% confident that instead of getting 10 correct answers, he will range anywhere from 6 to 14 correct answers. Problem 17A addresses the question of Statistical Confidence.

This probability problem is a humorous example of a Binomial Distribution, which occurs when there is a collection of events, each of which can only have one of two outcomes. (In Dewey's case the outcomes are his answering questions either correctly or incorrectly.) A real-world, important application of the Binomial Distribution principle is found in the field of population genetics.

If a gene has two variants (called "alleles") and the probability of each is p and q, respectively, then $p + q = 1$ (and $q = 1 - p$ as in Dewey's problem). Mating of two individuals can result in the three combinations of pp, pq and qq. If p is dominant and q is recessive, individuals with combination pp show the dominant trait, those with pq show the dominant trait but are carriers of the recessive trait, and those with qq show the recessive trait.

Over a large population, the probability of these combinations occurring can be calculated using the identity, $(p + q)^2 = 1$. The result is the Hardy-Weinberg equation: $p^2 + 2pq + q^2 = 1$, where p^2 is the probability that the pp combination will occur, 2pq is the probability that the pq combination will occur and q^2 is the probability that the qq combination will occur.

This equation has been used extensively to study population genetics and to determine if the gene pool of a population studied is in equilibrium or is still changing.

17A Confidence

Dewey has also just taken a nationwide exam in basic math skills along with 100,000 other students. On this exam, the questions are graded so that any score at all is possible between 0% and 100%. Here, Dewey was more prepared and received a grade of 70%. Dewey, figuring that he is really an average kind of guy, sitting atop the bell curve, believes that the national average is also 70%.

What is the required sample size of a survey, whose average score is 70%, to assure Dewey, to within 95% confidence level, that the national average is between 67% and 73%?

How would the required sample size change if Dewey had had a score of 80% and sought 95% confidence that the range was 77% to 83%? If his score were 90% (seeking 95% confidence in the 87%-93% range)?

Solution

Because we have a very large number of students taking the test, we can use the Normal Distribution as the result for the total population of 100,000 students.

To achieve a 95% confidence level, scores must be between -1.96 standard deviations and +1.96 standard deviations of the mean. That is 95% of the bell curve lies between $\mu - 1.96\sigma$ and $\mu + 1.96\sigma$, where σ is the standard deviation of the sample population.

If the mean of the sample population*, $\hat{p}$, is a probability of .70 (that a student will answer a question correctly), then the desired range from 67% to 73% represents a confidence interval of $\pm.03$.

Setting $.03 = 1.96\sigma$ yields $\sigma = .0153$.

The relationship between σ, $\hat{p}$ and n, the required sample size, is

$$\sigma = \sqrt{\frac{\hat{p}(1-\hat{p})}{n}} .$$

Solving for n yields

$$n = \frac{\hat{p}(1-\hat{p})}{\sigma^2}$$

For $\hat{p} = .7$ and $1 - \hat{p} = .3$,

$$n = \frac{(.7)(.3)}{(.0153)^2} = 897 \text{ students.}$$

If Dewey's score had been 80%, $\hat{p} = 0.80$ and $1 - \hat{p} = 0.2$.

Then,

$$n = \frac{(.8)(.2)}{(.0153)^2} = 684 \text{ students.}$$

And if Dewey's score had been 90%, $\hat{p} = 0.90$ and $1 - \hat{p} = 0.1$.

Then,

$$n = \frac{(.9)(.1)}{(.0153)^2} = 385 \text{ students.}$$

* By convention, the circumflex, or the "hat," above a parameter symbol means that it refers to the sample being measured. The same symbol without a hat refers to the entire population.

The advent of statistics courses in high school is a relatively new development. In the mid twentieth century, there were a few courses, at most, nationwide. Today, virtually every high school offers them. Why is this?

I think there are two answers to this question. The first is that statistics has become a more familiar and more integral part of work and social life in the information age society.

Numbers, in statistical format, are commonly used in virtually all arenas of business and "stats" has become an accepted norm for discussing almost any topic including sports, leisure, investing, etc. Perhaps the ultimate application of statistics in the job marketplace is doing insurance actuarial work. Market researchers also use statistics extensively.

Second, educators seek to reflect in their curricula those areas that will benefit students greatly. Statistics certainly is one such area. Also key in the decision to put Statistics into high school is that the introductory course requires skills that are commensurate with the rest of high school studies.

Of course, "Stats" can become very complex, which is probably why it started out as a college course and then worked its way down.

This problem makes use of the natural fact that when large numbers of cases are involved, the Normal Distribution comes into play. This distribution may be as central to Statistics as the Pythagorean Theorem is to geometry (as we have seen). It is from this distribution that we can get a sense of what we can be "sure" about.

As in Problem 16A, an application of this statistical principle can be found in the field of genetics. Quantitative characteristics are those that vary in the population in tiny gradations over a wide range. This is caused by polygenic inheritance which means that several genes are affecting the characteristic.

An example of this is human skin color which is determined by at least three genes. If each gene has a dark skin allele and a light skin allele, the results for the population are a histogram of skin tones that approximate a bell curve as the number of genes involved increase. Environmental factors such as exposure to the sun also smoothes out the bell curve.

In Dewey's case, I like the part about how Dewey thinks he represents everyone else: that he is the mean. Well, when you have one sample, assuming it is the mean is the best approximation you can make. But, generalizing from one example usually leads to trouble.

More seriously, though, the problem offers a set of deeper philosophical and personal questions. What do I know? What can I know? How sure can I be? What does "sure" mean to me? And then, it can become more practical, as in How much can I invest to find out the answers?

The principles underlying this problem can answer some of these questions quantitatively. We choose our confidence limits and all the rest flows from there. Are we willing to pay for 99% confidence or will be able to sleep well at 95%? Can we trust 90%?

18A Correlation

It is now the end of the semester and Dewey is reflecting on his math class achievements. He studies the grades he received on the exams that were given twice a month and wants to see if he can learn something about himself. He would like to explore whether there were any outside (of math class) influences that affected his performance on the exams.

After pondering, Dewey concludes that there are three things that worry him on a daily basis: dealing with cold weather, running out of money and keeping up a good relationship with his girlfriend who is at another school. Could any of these factors be influencing his performance on math exams?

To see if any of these are possible, Dewey does some research from the weather service web site, his bank records and his telephone bills. He compiles the following table.

Date of Math Exam	Test Score on Math Exam	Temperature at 8:00 am	Account Balance	Phone Minutes Previous Night
Feb 1	92	22	450	28
Feb 15	85	19	475	25
Mar 1	98	21	275	32
Mar 15	81	29	325	20
Apr 1	86	41	375	30
Apr 15	88	38	300	29
May 1	90	50	400	31
May 15	95	52	410	35
Jun 1	94	48	390	35
Jun 15	93	73	380	35

How well correlated are the temperature, his account balance and his phone time, individually, with his math exam scores?

Are there any improvements that Dewey can make next year based on this information?

Solution

The correlation coefficient, r, of two random variables measures how close to a straight line their scatterplot would be. The closer to a straight line it is, the stronger the linear relationship between the variables. The possible range of r is -1 to 1. An r value of -1 means the variables are linearly related but that one variable is increasing while the other is decreasing. An r value of 1 means the two variables move together in the same direction in a perfectly linear relationship. An r value of 0 means there is absolutely no correlation between the two variables.

The formula to calculate r, the correlation coefficient, is

$$r = \frac{\overline{xy} - \overline{x} \cdot \overline{y}}{\sqrt{\left(\overline{x^2} - \overline{x}^2\right)\left(\overline{y^2} - \overline{y}^2\right)}}$$

where x and y are the two variables and a bar across the top of a variable's symbol denotes the average of that variable from the data provided.

There are three cases to examine.

First, x represents the exam grades and y represents temperature.

A table of calculations shows:

Date	x	y	xy	x^2	y^2
2/1	92	22	2024	8464	484
2/15	85	19	1615	7225	361
3/1	98	21	2058	9604	441
3/15	81	29	2349	6561	841
4/1	86	41	3526	7396	1681
4/15	88	38	3344	7744	1444
5/1	90	50	4500	8100	2500
5/15	95	52	4940	9025	2704
6/1	94	48	4512	8836	2304
6/15	93	73	6789	8649	5329
Average	90.2	39.3	3565.8	8160.4	1808.9

Then,

$$r = \frac{3565.7 - (90.2)(39.3)}{\sqrt{(8160.4 - 90.2^2)(1808.9 - 39.3^2)}} = 0.260$$

Second, x represents the math grades and y represents account savings

Date	x	y	xy	x^2	y^2
2/1	92	450	41400	8464	202500
2/15	95	475	40375	7225	225625
3/1	98	275	26950	9604	75625
3/15	81	325	26325	6561	105625
4/1	86	375	32250	7396	140625
4/15	88	300	26400	7744	90000
5/1	90	400	36000	8100	160000
5/15	95	410	38950	9025	168100
6/1	94	390	36660	8836	152100
6/15	93	380	35340	8649	144400
Average	90.2	378	34065	8160.4	146460

Then,

$$r = \frac{34065 - (90.2)(378)}{\sqrt{(8160.4 - 90.2^2)(146460 - 378^2)}} = -0.104$$

Third, x represents the math grades and y represents minutes of telephone time

Date	x	y	xy	x^2	y^2
2/1	92	28	2576	8464	784
2/15	85	25	2521	7225	625
3/1	98	32	3132	9604	1024
3/15	81	20	1620	6561	400
4/1	86	30	2580	7396	900
4/15	88	29	2552	7744	841
5/1	90	31	2790	8100	961
5/15	95	35	3325	9025	1225
6/1	94	35	3290	8836	1225
6/15	93	35	3255	8649	1225
Average	90.2	30	2724.9	8160.4	921

Then,

$$r = \frac{2724.0 - (90.2)(30)}{\sqrt{(8160.4 - 90.2^2)(921 - 30^2)}} = 0.836$$

Correlation with temperature is weak; wear an extra sweater.
Correlation with account balance is virtually non-existent; don't worry about it.
Correlation with phone time is quite strong; keep that relationship going with more phone time.

Correlation is a fascinating subject because we, as humans, can try to correlate anything. Stock market investors have correlated the ups and downs of the New York Stock Exchange based on the winning conference of the Super Bowl. Others have tied the market to the level of Lake Erie... or was it one of the other Great Lakes? No matter.

The great part of this problem is that it demonstrates a straightforward way to show linear correlation. That's the way most people think... linearly. Actually, we could have looked for correlation using any other function such as higher order polynomials, exponential functions, trigonometric functions, etc. For now, though, we're going with linear.

Even if two variables are totally correlated with r = 1, it is possible that there is an underlying reason that is affecting both of them. That is, the cause and effect of one on the other may be nil. Rather, something else is making them do what they do and seems to be putting them in sync. This is an important consideration when applying the results of a calculation, such as the one for r. Still, Dewey will be well served by wearing a sweater, not fussing over his money and building an ongoing relationship with someone who likes spending phone time with him.

Another important perspective on this problem is the calculation of the values used to determine r. As shown, tables can be constructed to provide the intermediary parameter values for the final calculation. A calculator is certainly a welcome friend to this task. Better yet, a spreadsheet can be set up to do the calculations once the formulas have been programmed in. More so, a calculator will calculate all the interim parameters, the final value of r and a whole lot more with a couple of button pushes after the list of data has been entered.

Here is the big question. Will students learn statistics by knowing how to operate a calculator? Some yes, maybe most no. Perhaps the answer is to have students work some problems "long hand" once in a while so they get to reinforce the theory. Then, they can do the large work load using the calculator as a tool. And, even with a calculator, oversight is critical because errors can seep in and an experienced eye on the final answer is needed to assure accuracy.

19A   Arc Length

What is the length of the curve: $y = x^{3/2}$ from $x = 0$ to $x = a$?

Solution

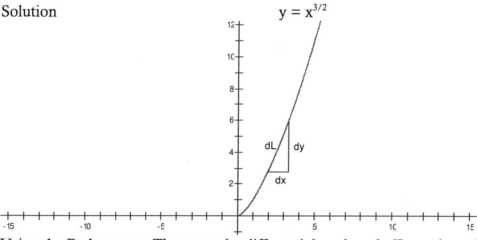

$y = x^{3/2}$

Using the Pythagorean Theorem, the differential arc length dL can be written as:

$(dL)^2 = (dx)^2 + (dy)^2.$ Then $dL = \sqrt{1+(\dfrac{dy}{dx})^2}\,dx$ and

$L = \displaystyle\int_0^a \sqrt{1+(\dfrac{dy}{dx})^2}\,dx$ which is the general formula for arc length.

For $y = x^{3/2}$, $\dfrac{dy}{dx} = \dfrac{3}{2}x^{1/2}$ and

$\left(\dfrac{dy}{dx}\right)^2 = \dfrac{9}{4}x$

Then, $L = \displaystyle\int_0^a \sqrt{1+\dfrac{9}{4}x}\,dx$

To integrate, use the "Substitution Method" where

$u = 1 + \dfrac{9}{4}x$ and $du = \dfrac{9}{4}dx$

Then, $L = \displaystyle\int \sqrt{u}\,\dfrac{4}{9}\,du = (\dfrac{4}{9})u^{3/2}(\dfrac{2}{3}) = \dfrac{8}{27}u^{3/2}$ and

$L = \dfrac{8}{27}(1+\dfrac{9}{4}x)^{3/2}\bigg]_0^a$

$L = \dfrac{8}{27}\left[(1+\dfrac{9}{4}a)^{3/2} - 1\right]$

$L = \dfrac{(4+9a)^{3/2}-8}{27}$

142

The derivation of the formula for arc length is usually included in most calculus texts in the chapter(s) that also address the area under a curve, the surface area of revolution and the volume of revolution. The striking difference about the arc length section is that there are so few examples given in the text, both in the narrative and in the exercises.

The reason for this is that there are very few "stand-alone" functions, $y = f(x)$, for which $\sqrt{1+(\frac{dy}{dx})^2}$ is integrable. That is why $y = x^{3/2}$ is almost always used as the example. One example of this function in nature is Kepler's Third Law, where the time for a planet to orbit the sun, T, is proportional to $r^{3/2}$, where r is the distance from the sun. Since the integral can be calculated in closed form for Kepler's Third Law, it is used in solving problems in astronomy, astronavigation, satellite tracking, etc.

Another "stand-alone" function is $y = \frac{1}{2}x^2$. Then, $\frac{dy}{dx} = x$ and

$$\int\sqrt{1+\left(\frac{dy}{dx}\right)^2}dx = \int\sqrt{1+x^2}dx = \frac{1}{2}[x\sqrt{1+x^2}+\ln(x+\sqrt{1+x^2})], \text{ which is fairly complex.}$$

There are some "combination" functions that also work, such as $y = \frac{x^3}{3}+\frac{1}{4x}$. Then

$$\frac{dy}{dx} = x^2 - \frac{1}{4x^2} \text{ so } \left(\frac{dy}{dx}\right)^2 = x^4 - \frac{1}{2}+\frac{1}{16x^4} \text{ and}$$

$$\sqrt{1+(\frac{dy}{dx})^2} = x^2 + \frac{1}{4x^2} \text{ which can be easily integrated to } \frac{x^3}{3}-\frac{1}{12x^3}.$$

Over the years, most math tests just require the student to "set up" the integral solution without solving it. It is not surprising that this formula gets as little attention as it does.

On the other hand, calculators can work out definite integrals even when the integrand is not integrable in closed form. This would be very handy for most real-world arc length problems.

Recently, I revisited a problem I remembered working in my advanced calculus class: determining the shortest distance between two quadratic curves. The closed form solution involves partial derivatives and the use of Lagrange multipliers. It is an elegant and satisfying use of advanced principles. Then, I found I could get a very accurate estimate of the solution with a few button pushes of a calculator. The bottom line is that calculators won't teach math principles well but they will save computation time and will get good approximate answers when closed form solutions are not available.

20A   Surface and Volume of a Solid of Revolution

What is the surface area and volume of the solid defined by the hyperbola
$y = \dfrac{1}{x}$, for $0 < x \le 1$, revolved around the y − axis?

Can this three dimensional figure be painted?

Solution

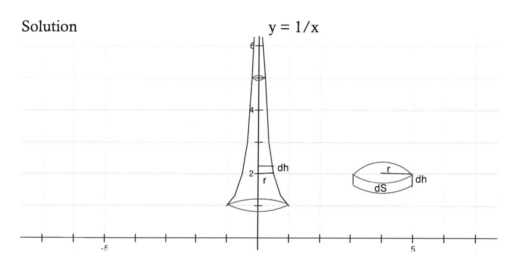

y = 1/x

The surface area of an open cylinder is $S = 2\pi rh$ where r is the radius and h is the height.

The differential increment in surface area is $dS = 2\pi r\,dh$

Here,   r = x and dh = dy.   Then $S = 2\pi \displaystyle\int_{1}^{\infty} x\,dy$

But $x = \dfrac{1}{y}$, so the integral becomes

$S = 2\pi \displaystyle\int_{1}^{\infty} \dfrac{1}{y}dy = 2\pi \ln y\big]_{1}^{\infty} = \infty$

The surface area of this solid of revolution is infinite and therefore cannot be painted.

The volume of a cylinder is $V = \pi r^2 h$ where r is the radius and h is the height.
The differential increment of volume is $dV = \pi r^2 dh$.

Here, r = x and dh = dy.

Therefore the volume of the solid of revolution is:

$V = \pi \displaystyle\int_{1}^{\infty} x^2 dy$ but $x^2 = \dfrac{1}{y^2}$ so $V = \pi \displaystyle\int_{1}^{\infty} \dfrac{1}{y^2}dy = \dfrac{-\pi}{y}\Bigg]_{1}^{\infty} = \pi$

The volume of this solid of revolution is finite (π) and so it can be filled with paint.

144

Solving for the surface area of a solid of revolution is a practical approach to figuring the amount of material needed for a manufactured product. Finding the volume also is important, especially when the packaging must declare the volume of the contents. Although figuring out surface and volume can be done by trial and error, and/or after the fact measuring, modern design and cost analysis methods calculate these parameters using this method.

The answer is that since the surface area is infinite, an infinite amount of paint would be needed and, therefore, it could not be painted. Not being able to paint the solid because it has infinite surface area is a surprising result considering that the volume of the solid is finite. How can you have a container with a finite volume and an infinite surface area? Wouldn't filling the volume with paint, in fact, paint the entire (interior) surface? How can this paradox be resolved? These kinds of questions can lead the student into many fertile fields of study such as limits, series and sequences, fractals, topology and many more.

When teaching about conic sections, I like to include a discussion of the origins of their names. The word hyperbola comes from the Greek "hyperbole" which means "a throwing beyond." Class discussion about what is thrown where can liven up a class for sure. Then, throw into the mix, the other English word that comes from hyperbole, which is "hyperbole" and we get even more fun. "Hyperbole" in English means "an obvious exaggeration, used for effect," usually in a rhetorical setting. Well, that is certainly a "throwing beyond." All this can generate a lot of laughs.

Similarly the word parabola comes from the Greek "parabole" which means "a placing beside" or "comparison." Again, there is another English word from this root which is a "parable," a short allegorical story. And, again, the word ellipse comes from the Greek "ellipsis" which means "omission." The counterpoint English word here is also "ellipsis" which refers to the "..." we use in writing to show something is missing. Note that in English, ellipse and ellipsis both have the same plural spelling of "ellipses."

Another interesting word is "frustum" which comes from the Latin word "piece." It is the shape of the base of a cone with the top cut off. For example, when two parallel planes intersect a cone perpendicular to the axis of the cone, two circles (of different sizes) are the conic sections and a frustum of the cone is formed. These are handy shapes used in circuses as pedestals for the wild animal acts. When you get several of these together, we use the plural "frusta." If you are teaching a math class, see what response you get if you ask them to bring in two cone frusta.

Another good addition to "cutting up" on conic sections is to study loci. The easiest is a circle, which is the locus of points equidistant from one point; followed by a parabola, which is the locus of points equidistant from a point and a line. An ellipse is the locus of points, the sum of whose distances from two points is a constant; and a hyperbola has a constant difference of distances between a point and a line.

21A   Two Intersecting Cylinders

What is the volume of intersection between two right cylinders, both of radius r, whose axes intersect perpendicularly?

Solution

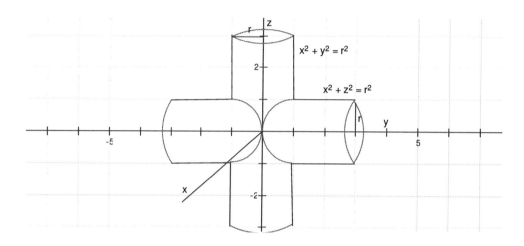

The diagram above shows an overview of the intersection of the two cylinders.

The diagram below shows a cutaway view of the intersection in the first octant.

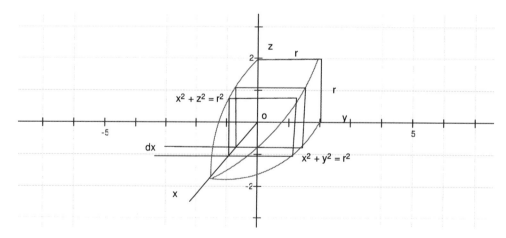

The equations of the two cylinders, as shown in the diagram, are:

$$x^2 + z^2 = r^2 \quad \text{and} \quad x^2 + y^2 = r^2$$

Therefore, $z = \sqrt{r^2 - x^2}$ and $y = \sqrt{r^2 - x^2}$

For each of the eight octants of the intersected volume,

$$dV = yzdx \text{ (as shown in the figure). Then,} \quad V = \int_0^r yzdx$$

Therefore, the entire volume of intersection is

$$V = 8\int_0^r \sqrt{r^2 - x^2}\sqrt{r^2 - x^2}dx \quad \text{Then}$$

$$V = 8\int_0^r (r^2 - x^2)dx$$

$$V = 8\left(r^2 x - \frac{x^3}{3}\right)\Bigg]_0^r = \frac{16r^3}{3}$$

This great problem has been in calculus texts for a long time. One of its interesting features is that although most students find it difficult to visualize what the intersection looks like, and they have difficulty drawing it, the math really does all the work. Once the equations are set up, the answer flows easily. Although the solution shown uses single integrals, there are corresponding solutions using double and triple integrals.

There are a couple of interesting aspects of the answer of $16r^3/3$ which are worth further exploration.

The first is that there is no factor of π in the answer. When was the last time you calculated the volume of any curved space such as a sphere, cylinder, cone, etc that did not have a factor of π in the answer?

The second is to consider that if we had used two square prisms (instead of circular cylinders), each with side 2r, the resultant volume of intersection would have been $8r^3$. Therefore the volume of the intersecting cylinders is 2/3 of the volume of the intersecting prisms.

Students can pursue these questions for additional insights. Are these two aspects related? Is there a geometric interpretation? What else can flow from this study?

22A Integration

Find $\int (e^{ax} \cos bx)dx$

Solution

To integrate this function, we must use the method called Integration by Parts.

We make a substitution of u for part of the integrand and dv for the other part, and we apply them to the identity:

$$\int u\,dv = uv - \int v\,du$$

In this case, let $u = e^{ax}$ and $dv = (\cos bx)dx$

We can differentiate u to get $du = ae^{ax}dx$

And we can integrate dv to get $v = \dfrac{1}{b}\sin bx$

Substituting yields

$$\int (e^{ax} \cos bx)dx = \frac{e^{ax}\sin bx}{b} - \frac{a}{b}\int (e^{ax}\sin bx)dx$$

Using the method of integration by parts, again, to evaluate the resultant integral, we use
$U = e^{ax}$ and $dV = (\sin bx)dx$

Differentiating U and integrating dV yields
$dU = ae^{ax}dx$ and $V = -\dfrac{1}{b}\cos bx$

Substituting back into the original equation gives us

$$\int (e^{ax} \cos x)dx = \frac{e^{ax}\sin bx}{b} - \frac{a}{b}\left[-\frac{e^{ax}\cos bx}{b} + \frac{a}{b}\int (e^{ax}\cos bx)dx \right]$$

We see that the new resultant integral is the same one we started with. We can now treat this integral as any algebraic expression and collect its like terms. The answer, then, is

$$\int (e^{ax} \cos bx)dx = e^{ax}\left(\frac{b\sin bx + a\cos bx}{a^2 + b^2} \right) + C \quad \text{where C is the constant of integration.}$$

Integration is the second of the two pillars of calculus; the first is differentiation. Integration is sometimes called "anti-differentiation." Being able to calculate or "accumulate" the area under a curve is an essential element in the study of calculus.

Romans had a hard time doing math because their number system was based on their alphabet. The more practical approach to figuring things out was to use a wooden board with stones which was operated similarly to an abacus. The word for stone, or pebble, in Latin is "calculus," which is where we get our word "calculate." ("Calculus" is also still used today for the stony build-up on our teeth.) The inventors, or discoverers, of calculus were Leibnitz and Newton who reused the word for this new mathematical dimension. By then, of course, they were using Arabic numerals for calculations because al Kwarizmi's book had been translated and the new system had been popularized by earlier European mathematicians such as Leonardo da Pisa, also known as Fibonacci (see Problem 1B.)

Integration by substitution and integration by parts are the two primary methods of integration when the integrand is not immediately integrable. The substitution method, if it works, requires only a single iteration, but integrating by parts can take one or more iterations.

The function $e^{ax} \cos bx$ is special in that the result of the first integration by parts yields an integral that requires integration by parts again; the result of which is to wind up back where we started, with the original integral. Then we can solve algebraically, as was done. This situation is very unusual and, in a way, makes it feel as if we did not really integrate the function in the sense that most other problems actually result in a direct answer. Folding back over oneself to the beginning and finding the solution through how the integral relates to itself is, indeed, a special case.

On the other hand, I find that if my first integration by parts has yielded another integral which requires integration by parts, I always hope that I will wind up with the original integral on the second round. If so, I know I will be able to solve for the integral in closed form. If the second resultant integral is yet again some different integral, it is time to start worrying that integration by parts may not work at all but, instead, will lead to an infinite iteration of increasingly more difficult integrands to integrate.

23A Related Rates

A boy is standing b feet from the railroad track, and follows an oncoming train, traveling at v feet/sec, by turning his head. How fast will his head be turning when the train passes directly in front of him?

Solution

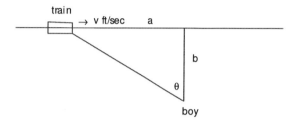

From the diagram,

$\tan \theta = \dfrac{a}{b}$.   Then

$a = b \tan \theta$

We now differentiate with respect to time, t, and get

$$\frac{da}{dt} = b \sec^2 \theta \frac{d\theta}{dt} + \tan \theta \frac{db}{dt}$$

But $\dfrac{da}{dt} = v$ (as given) and $\dfrac{db}{dt} = 0$ because the boy is standing still.

Substituting yields

$$v = b \sec^2 \theta \frac{d\theta}{dt} \text{ and}$$

$$\frac{d\theta}{dt} = \frac{v}{b \sec^2 \theta}$$

When the train is directly in front of the boy, $\theta = 0$ and $\sec^2 \theta = 1$, so

$$\frac{d\theta}{dt} = \frac{v}{b} \text{ rad/sec}$$

As mentioned in the comments of the last problem, differentiation is the first pillar of calculus. It is a descriptor of instantaneous change. And since nature is changing constantly, differentiation can be used to model almost all of life. Typically, modern technology uses differential equations, which describe the relationship between derivatives, to model situations integration techniques are then used to solve. This approach enables us to understand the world around us and to create products and processes for the common good. Examples of success include the arenas of mechanics, heat, sound, electricity, magnetism, optics, atomic physics, chemistry, biology, medicine, the arts, finance, risk analysis and many more.

Related rate problems are a practical application of the study of calculus. Texts usually have many examples to be worked and are often posed as word problems (as was the case here).

Here, the problem is stated as the "general case" of a train approaching but, usually, calculus texts have innumerable variants of particular cases with numerical values specified.

All calculus texts teach how to do related rate problems. The simplest way to approach them is to use a four step process which works consistently well.

The first step is to draw a picture of the situation and to label all the parameters given and desired. This may be the most difficult step because it requires understanding the problem as a word problem and visualizing the scenario. The vast majority of related rate problems when drawn will be a triangle (as was the case here).

The second step is to write an equation that describes the relationship between the parameters in the diagram. Usually, a simple trigonometric function or the Pythagorean Theorem is all that is needed.

The third step is to differentiate the equation with respect to the variable required in the problem. Again, for the vast majority of cases, this will be time t. This differentiation will often be an implicit differentiation and will require use of the Chain Rule to differentiate correctly.

The fourth step is to substitute in the values of the parameters given for the time in question. It is best not to look at the specific data given until step four is reached because focusing on the parameter value details at the beginning is only a distraction.

This outline is, in fact, a specific example of an approach that can be used for word problems, in general. This general approach could be stated as: draw a picture and label the parameters, write an equation that represents the picture, solve the equation and substitute in the values that are known.

Extrapolating even further gives us a useful formula or process for life in a larger sense. Know who you are and with whom you are dealing. Identify the environment and its elements in which you are working. Figure out what you need to do to find the best solution for all involved.

About the Author

Lewis Forsheit grew up in Brooklyn, New York, where he graduated from Abraham Lincoln High School and was Co-Captain of the Math Team in 1959. After receiving a BS and MS in Electrical Engineering from Columbia University School of Engineering, he was a successful aerospace engineer and engineering manager specializing in aircraft and booster rocket control systems. (Yes, he really was a rocket scientist.) In 1977 he changed careers to become a sales account executive and sales manager in the computer industry. Upon retiring from both careers in 2002, he returned to his academic passion of mathematics and became a high school math teacher, thereby fulfilling his destiny by joining the ranks of generations of teachers in his extended family. He teaches mathematics at Wildwood School and lives in Los Angeles with his wife, who (by his own estimation) is the world's greatest biology teacher. Together, they enjoy spending time with the families of their two married daughters.

The author encourages readers to comment on any aspect of the book via email: lew_forsheit@earthlink.net.

Printed in the United States
By Bookmasters